From Oracle Bones to Computers

From Oracle Bones to Computers

The Emergence of Writing Technologies in China

Baotong Gu

Parlor Press
West Lafayette, Indiana
www.parlorpress.com

Parlor Press LLC, West Lafayette, Indiana 47906

© 2009 by Parlor Press
All rights reserved.
Printed in the United States of America

SAN: 254-8879

Library of Congress Cataloging-in-Publication Data

Gu, Baotong, 1963-
 From oracle bones to computers : the emergence of writing technologies in China / Baotong Gu.
    p. cm.
  Includes bibliographical references and index.
  ISBN 978-1-60235-101-1 (acid-free paper) -- ISBN 978-1-60235-100-4 (pbk. : acid-free paper)
  1. Writing--China--Materials and instruments--History. 2. Written communication--Technological innovations--China--History. 3. Written communication--Technological innovations--History. I. Title.
  Z45.G83 2009
  303.48'30951--dc22
                                      2009008637

Cover design by David Blakesley.
Printed on acid-free paper.

Parlor Press, LLC is an independent publisher of scholarly and trade titles in print and multimedia formats. This book is available in paper, cloth and Adobe eBook formats from Parlor Press on the World Wide Web at http://www.parlorpress.com or through online and brick-and-mortar bookstores. For submission information or to find out about Parlor Press publications, write to Parlor Press, 816 Robinson St., West Lafayette, Indiana, 47906, or e-mail editor@parlorpress.com.

To my parents and billions of other Chinese who
never had the privilege to enjoy literacy

Contents

Acknowledgments	*ix*
1 Introduction: (De)Mystifying the Chinese Culture	*3*
2 (Un)loading Technology	*24*
3 Rhetoricizing and Operationalizing Technology	*68*
4 Oracle and Bronze Inscriptions	*103*
5 Early Forms of Pen, Ink, and Paper	*118*
6 The Modern Form of Paper	*136*
7 Block Printing and Movable Type	*151*
8 The Chinese Typewriter	*167*
9 The Computer and the Internet	*177*
10 Conclusion: Toward a More Pluralistic Model of Knowledge Construction	*231*
Appendix: Milestone First Events in China's Internet Use	*235*
References	*243*
Index	*257*
About the Author	*267*

Acknowledgments

It is not until you undertake a project like this that you realize how much you must rely on the selfless efforts of others. Though only my name appears on the cover as the author of this book, I can never justifiably claim sole authorship. I owe the completion of this book project to a number of people, to whom no language is adequate to express my deep-felt gratitude.

This book originated in my dissertation work at Purdue, so my first word of thanks goes to the members of my dissertation committee: Patricia Sullivan, James Porter, Janice Lauer, Johndan Johnson-Eilola, and Tony Silva. Their insightful comments and advice molded my initial thinking on the project and helped lay a solid foundation for this book.

I am also deeply grateful to my colleagues at Georgia State University, especially to Randy Malamud and George Pullman for their constant moral support and encouragement, which provided me with much of the strength and resolve I needed to complete the project; to my Rhetoric and Composition colleagues for their help and advice in many facets of my professional life, and to Angela Hall-Godsey for taking care of much of the drudgery work of Lower Division Studies so that I could concentrate on this book.

I owe a special thank-you to the staff at Parlor Press, especially to editor David Blakesley for steering me along the right course throughout this publication process and making it basically painless, to the two anonymous reviewers for their selfless work and very helpful advice on improving my manuscript, and to copyeditor Rebecca Longster for her meticulous attention to details and her overall excellent job, which has spared me a lot of embarrassment.

My final word of thanks goes to my family—my wife, Li, and my children, Shelly and Shawn—whose patience and moral support have made this project possible.

From Oracle Bones to Computers

1 Introduction: (De)Mystifying the Chinese Culture

> These "Ocean Men" [foreigners], as they are called . . . are tall beasts with deep sunken eyes and beaklike noses. The lower part of their faces, the backs of their hands, and I understand, their entire bodies are covered with a mat of curly hair, much as are the monkeys of the southern forests. But the strangest part about them is that, although undoubtedly men, they seem to possess none of the mental faculties of men. The most bestial of peasants is far more human, although these Ocean Men go from place to place with the self-reliance of a man of scholarship and are in some respects exceedingly clever. It is quite possible that they are susceptible to training and could with patience be taught the modes of conduct proper to a human being. (cited in Oliver, 1971, p. 5)

The above passage is the depiction by a scholarly Chinese of his first impression of white men in his country. The absurdity of such a perception is beyond question. However, that somebody should view a fellow human being, though of a different race, in such a manner is by no means a uniquely Chinese phenomenon. What is of interest here is not the absurdity of the story or the ignorance of a particular human being or species, for mere ignorance cannot adequately explain such a phenomenon. Nor is the observer's possible prejudice to take the blame. What is at work is the observer's enculturated consciousness, which is facilitating, albeit to a woeful effect, his reconstruction of the white men based on his experience of members of the white race, which, unfortunately, is extremely limited.

What, then, does this have to do with writing technologies?

The answer, which is far more complicated than I can articulate here, lies in people's perception and subsequent rhetorical construction of the meaning of a particular phenomenon, ideology, or artifact . . . be it a foreigner or a new technology. What a foreigner is, in the above case, depends on the observer's experience with the human species, which is limited to his own race, and makes sense only in the context of his own race. In a similar way, what a writing technology is—what it means, how it should be used, and how it should be further developed—depends, to a large extent, on how people (the participants of technology development) perceive and define this technology.

Therefore, it is the main goal of this book to interpret such perceptions and their implications on the development of technologies, particularly writing technologies. However, the purpose of this book is multifold: to define technology, technology development, and technology transfer; to present a coherent and comprehensive picture of the emergence of various writing technologies throughout the history of China, including, for example, oracle inscriptions, bronze inscriptions, pen, ink, early forms of paper (such as bamboo, wood, and silk), paper of the modern form, block and movable type printing, the Chinese typewriter, and the computer; to explore the impact of these writing technologies on the respective historical periods with special regard to writing and communication behavior; and, most importantly, to deconstruct the social, political, and cultural contexts and their shaping influence on writing technology developments.

CHINA: AN INTRIGUING CASE

Several factors make China an intriguing case for the study of writing technology development: a long history of civilization (over five thousand years), a source of some of the most important inventions of writing technologies (such as pen, ink, paper, and printing), a confusing pattern of development (with some periods flourishing with milestone emergences and others completely void of development), a combination of both native developments and foreign transfers (thus rendering itself an appropriate case study of technology transfer), and an intriguing, if not mysterious, writing system.

The first intriguing factor, China's long civilization of over five thousand years, provides a rich site for any study, not the least of which is writing technology development. The history of the written Chinese itself can be traced as far back as the Dawenkou culture between the twenty-eighth century BCE and the twenty-third century BCE (Peng et al., 1989, p. 432). Over four or five thousand years of evolution, the Chinese civilization underwent a series of differing cultures during different historical periods and dynasties. This myriad of cultures adds an intriguing complexity to the context of writing technology development. In addition, a mature civilization as the site of study yields validity to our investigation, which targets culture as the scene of development.

A second intriguing factor lies in the fact that the Chinese civilization has been host to several of the most important inventions in writing technologies, such as pen, paper, ink, and printing. As early as the sixteenth century BCE, in the Shang Dynasty, when there was hardly any writing, not to mention writing technology, in other parts of the world, the Chinese were already using turtle shells and other animal bones for oracle inscriptions (Xia et al., 1979, p. 1673). This was arguably the earliest form of writing technology in the history of human civilization.

Another notable era of writing technology development is the subsequent period of the Zhou Dynasty (from the eleventh to the third century BCE). This so-called classical period of China saw the use of bamboo pens, soot ink, and bamboo and wood slips as primitive forms of paper (Carter, 1955, p. 94). Then came the invention of the writing brush, made of hair, in the third century BCE. This invention, according to Thomas Carter, "worked a transformation in writing materials, [which was] indicated by two changes in the language," one being that "the word for chapter used after this time means 'roll'" and the other being that "the word for writing materials becomes 'bamboo and silk' instead of 'bamboo and wood'" (1955, p. 94).

China's pioneering role in the invention of writing technology was best evidenced by two major inventions in the later history of China: paper and printing. Paper, which was commonly believed to have been invented around 105 CE (a claim disputed by many scholars), liberated the Chinese from the heavy bamboo "papers." Printing in China witnessed two landmark inventions: that of block printing and that of the movable type printing. Block printing was invented in the golden age of literary and artistic prosperity in the eighth century, which saw the birth of some of the greatest poets and artists in China's history. Then in the eleventh century came the invention of movable type printing.

These two inventions of printing quickly spread to European countries and revolutionized, to a large degree, the writing technologies of their respective historical periods.

Although China cannot claim credit for the modern forms of all these writing technologies, few civilizations parallel China in spearheading writing technology developments in earlier historical periods. This multiplicity of writing technology inventions yields rich options for meaningful studies of writing technology development.

A third intriguing factor about China is that the history of writing technology development exhibits a confusing pattern, with flourishing developments in some historical periods and total inertia in others. The modern period, which spans a term of several centuries, is surprisingly impoverished in writing technology inventions, when compared with the early part of Chinese history. The West, in the meantime, seems to have been taking big strides in advancing various writing technologies. In the second half of the nineteenth century, for example, driven by an ideology of systematic management, America invented and discovered the use of the typewriter, duplicating methods such as hand printing and press printing, and filing systems. These inventions furnished the necessary technological means for the then increasing demand for written communication "to provide consistency, exactness, and documentation" (Yates, 1989, p. 22). However, the culmination of the development of writing technologies in the West, so far, has to be the invention and use of the computer, which has come to pervade and revolutionize writing in almost all fields and disciplines. Conversely, with regard to this most recent, most revolutionizing writing technology, China is so far behind the West in its development and implementation as to strain credulity. Yet there seems to exist no theory that could adequately explain how a country that had pioneered most of the way in advancing writing technologies in the history of human civilization could be so outpaced in its computer technology. Economic underdevelopment, though a seemingly feasible explanation for such a phenomenon, can only be an oversimplified justification. The answer has to lie in the more complicated cultural context that provides nourishing ground for the growth of new technologies. This idea alone adds interesting facets to our study.

A fourth factor that adds an intriguing dimension to our study is that the Chinese history of writing technology development encompasses both native developments and foreign transfers. China's early

Introduction: (De)Mystifying the Chinese Culture 7

computer technology, for example, was a wholesale transfer from the West. Yet, due to the unique nature of Chinese script, the Western design of the computer was unusable for producing Chinese characters. Fundamental modifications, therefore, were incorporated into the design to localize this technology in China, to render it useful in the Chinese context. This act of transfer and localization is an important aspect not to be overlooked in any study of writing technologies.

A fifth, and final, intriguing factor about the China case is the unique nature of the Chinese language/script, which is fundamentally different from Roman-based scripts such as English. While English is a phoneme-based alphabetical language, Chinese is an ideogram-based script language. (A more elaborate discussion of the nature of Chinese script is provided in the next section.) Translated into computer terms, each English letter takes up one byte of space in the ASCII system whereas each Chinese character occupies two. Therefore, in localizing the computer technology for the Chinese context, the specific changes in the design of the computer to accommodate such differences add yet another interesting facet to the study. (In fact, the complexities of this script with regard to both its nature and history defy easy characterization and warrant a separate section with a more elaborate and in-depth discussion. This, however, is certainly not meant to undermine the significance of the other factors that have contributed to the intriguing nature of the China case.)

The Mystery of the Chinese Script

Although the nature of Chinese script (and the mystery surrounding it), as discussed in the preceding section, is only one of the five factors that make China's writing technology development an intriguing case, it deserves special treatment here for three distinct reasons. First, every writing technology in the history of China has been a direct outcome of the development of the Chinese language, the various stages of which demanded and dictated corresponding writing technologies that could accommodate the changing nature of the Chinese script. Second, the unique, complex nature of this script has had direct bearings and, in most cases, shaping influences on how the writing technologies were conceived and developed. Third, the Chinese language has undergone a long history of evolution (over 4,000 years) and many significant changes in the course of its development. Understanding this history is critical to a better understanding of writing technology development.

A meaningful starting point of discussion of the Chinese script, then, is its origin. Due to the long, pre-record history of the Chinese civilization, it is difficult, if not impossible, to pinpoint an exact historical point when Chinese script originated. This difficulty has been acknowledged by many researchers and is reflected in the differing periods they have designated for the origin of Chinese script (see, for example, Boltz, 1999; Cheung, 1983; "Chinese Language," 1997; Jian, 1979; Lattimore, 1946; Rodzinski, 1984). Various stories have been told about the origin of Chinese script, with many ancient ones pointing to a man named Cangjie:

> Cangjie, according to one legend, saw a divine being whose face had unusual features which looked like a picture of writings. In imitation of his image, Cangjie created the earliest written characters. After that, certain ancient accounts go on to say, millet rained from heaven and the spirits howled every night to lament the leakage of the divine secret of writing. Another story says that Cangjie saw the footprints of birds and beasts, which inspired him to create written characters. ("Chinese language," 1997)

The truthfulness of the stories is certainly questionable. More likely, Cangjie only sorted out the characters already invented by the people ("Chinese language"; Xia et al., 1979, p. 312). A recent discovery of some ancient tombs in Yanghe, Shandong Province, has unearthed a dozen pottery vessels dating back to a late period of the Dawenkou culture of about 4,500 years ago. Each of these pottery vessels bears a character, and these characters "are found to be stylized pictures of some physical objects" and are therefore called pictographs ("Chinese language," 1997). These pictographs are already quite close in style and structure to the oracle bone inscriptions of the sixteenth century BCE but predated the latter by about 1000 years ("Chinese language," 1997).

Similar discoveries have been made in the last few decades, with some unearthed, character-bearing pottery vessels dating back to as early as the period between 4800–4200 BCE (Cheung, 1983, p. 324–25). Of course, these characters are less than regular enough to form a systematic script, the first of which is usually considered to be the oracle bone inscriptions of the Shang Dynasty, around the sixteenth century BCE.

As mentioned earlier, unlike that of many European languages, Chinese script is not an alphabetic script, but a script of ideograms. According to Feibo Du (1998) and "Chinese language" (1997), the formation of Chinese characters follows three principles: hieroglyphics (the drawing of pictographs), associative compounds, and pictophonetics.

Hieroglyphics, probably the earliest method of forming Chinese characters, refers to the method of forming a character according to the actual form of the object the character refers to. For example, the character for the sun was written as ☉, for the moon 𝕯, for water 〵〵, and for cow ψ. The similarity between the character and the object it signifies is obvious. As this script evolved over the centuries, these pictographs gradually acquired a square shape, with some being simplified and others complicated, but overall regularized and systematized. Hieroglyphics provided the basis upon which subsequent methods of character formation were developed.

Though easy to understand, pictographs have a serious drawback: they cannot express abstract ideas. To make up for such a drawback, associative compounds were developed to form characters that combine two or more pictographic characters, each with a meaning of its own, to express abstract ideas. For example, the sun ☉ and the moon 𝕯 combined to form the character *ming*, ☉𝕯, meaning "bright." The sun placed over a line forms the character *dan*, ☉, meaning "morning" or "sunrise."

Neither pictographs nor associative compounds, however, indicate how the characters should be pronounced. Hence, the method *pictophonetics* was developed. Pictophonetics combines two elements: meaning and sound, in forming characters. For example, the character for "papa," combines the element *fu* for the meaning (father) and the element *ba* for the sound.

The significance of these three methods of creating script is that they represent some of the most common measures of developing a systematic script for the ancient Chinese language. Researchers often use them as evaluation criteria for determining whether a particular historical period possessed a systematic script. Of course, later versions of Chinese script, especially that of the current Chinese language, contain characters that do not fall into any of the three categories. This is why some researchers (e.g., Jian, 1979) have argued that modern Chinese script should contain six essential features (more details later in the book). Nevertheless, they are valuable tools for our examination of the development of writing and writing technologies, especially in the early stages of Chinese history.

The Mystery of the Dominant Chinese Ideologies

A crucial aspect to an in-depth comprehension of the Chinese culture is the understanding of its dominant ideologies, for the development of any writing technology is inevitably shaped by the ideologies of a particular culture. In the history of Chinese thought, there have existed many different ideologies, but my discussion will focus on the three most dominant ones: Confucianism, Taoism, and Buddhism, because these three have exerted the most influence on the Chinese culture, not only across all geographical regions but also across most historical periods. Other ideologies, such as Legalism, Marxism-Leninism, and Maoism, are more or less regional or ephemeral; discussion of these ideologies will, therefore, be done only when they apply to my discussion of the development of a particular writing technology in subsequent chapters.

Confucianism

Confucianism, probably the most influential ideology in the history of Chinese thought and therefore worthy of a detailed examination, originated mainly in the teachings of Confucius and his disciples. Confucius (551–479 BCE) was born around the late period of Spring and Autumn and the early period of Warring States. Throughout his life, he was mostly poor, untitled, and without official position. Probably because of this, he devoted his whole life to learning and teaching. According to *Ci Hai* (an encyclopedic dictionary of the Chinese language), he had as many as 300 disciples, about 70 of whom became famous (p. 1119). Confucius was a philosopher, a political scientist, an educationalist, and a social critic. His ideas are mostly preserved in the so-called "five classics," namely, *The Book of Songs, The Book of History, The Book of Rites, The Book of Change,* and *The Spring and Autumn Annals,* and four books (i.e., *The Great Learning* [or *Ethics and Politics*], the *The Golden Medium* [or *The Book of Mean,* or *Central*

Harmony], *The Analects* [or *The Sayings of Confucius*], and *The Book of Mencius*). The book that most directly records his sayings is *The Analects.* Confucius's time was an era of instability. It was a time when the objective traditions of the land were being eroded by the influence of a subjective sophistry similar to that in the Greek tradition. "And it was Confucius who inspired a defense against these sophistic innovations by reasserting confidence in old principles and practices" (Ware, 1955, p. 10). Robert Oliver (1971) argues that Confucius set for himself the goal "to change the nature of Chinese civilization with a bloodless revolution" (p. 121). He may be overstating the case here, and such an assertion is obviously arguable, for it was never Confucius's intention to upset but to preserve the great tradition in Chinese civilization, or more accurately, as Thomas Cleary (1991) has asserted, to revitalize the culture "in its role as a means of cultivating human feelings and maintaining the integrity and well-being of a people" (p. 1).

Confucius's philosophy was deeply rooted in a concept of social order and harmony. He sought a society of harmony by means of self-purification by individuals, which was to be achieved through increased knowledge, for, "like Socrates, Confucius believed profoundly that one could not renounce what he knew to be right" (Oliver, 1971, p. 132). So, in essence, "Confucianism stood for a rational social order through the ethical approach, based on personal cultivation. It aimed at political order by laying the basis for it in a moral order, and it sought political harmony by trying to achieve the moral harmony in man himself" (Lin, 1938, p. 6). Confucius believed that the cultivation of the self would lead to the regulation of family life, which in turn would lead to the ordering of a national life. Therefore, one major means for attaining such a moral social order is through the education of the individual. Education is for the general enhancement of the individual and the success of groups—family, community, nation—to which the individual belongs (Cleary, 1991, p. 1–2). As Confucius said, "a piece of jade cannot become an object of art without chiseling, and a man cannot come to know the moral law without education" (Lin, 1938, p. 241). Once this piece of jade is "chiseled" and becomes a piece of art, it can help chisel others. However, as Mencius, the most faithful follower and developer of Confucianism, said, "Never has a man who has bent himself been able to make others straight" (Oliver, 1971, p. 169). So, the cultivation of the individual will leads to the cultivation of the family, and then of the community, and then of the nation, until finally we have achieved a moral social order.

However, cultivation of the individual must be based on the moral virtue of the humanness or humanity of human beings, for it is the moral foundation of social order. Though Confucius never clearly defined humanity, his concept of humanity can be understood in social terms: "being respectful at home, serious at work, and faithful in human relations" (Cleary, 1992, p. 3). Cleary identifies five characteristics in Confucius's conception of humanity, namely, respectfulness, magnanimity, truthfulness, acuity, and generosity (p. 4). Confucius believed that the measure of man is man. The whole philosophy of ritual and music, which Confucius emphasized in his writings as a part of the social order, is but to set the human heart right (Lin, 1938, p. 13).

An important part of this humanism is the concept of *jen*, variously translated as human, humane, humanitarian, humanity, kindness, benevolence, and true manhood. He considered it the highest human attainment "to find the central clue to our moral being which unites us to the universal order (or to attain central harmony)" (Lin, 1938, p. 185). To Confucius, when a man seeks to establish himself, he establishes others; when he wants to succeed himself, he helps others to succeed. Such a notion of self-improvement and social action is closely related to the notion of *jen*.

Another important concept in Confucius's conception of social order is *yi* (or justice, or duty, or principle). Although Confucius's notion of duty may seem to some people to be referring to an unquestioning obedience to superior authority, he never meant it to be obedience to dictators or rulers who pretended to advocate justice but really sought profit and advantage. Instead, it refers to an obligation to justice that will only strengthen the moral fiber of society.

One more concept in Confucius's notion of a moral social order is *li*, or known in varied translations as etiquette, propriety, or moral discipline. According to Confucius, the meaning of etiquette includes "concepts of mannerly behavior in day-to-day life, proper enactment of social rituals like marriage and mourning, and protocols for international and official occasions" (Cleary, 1992, p. 5). Lin (1938) also sees its close link with social practices and sees it as including folkways, religious customs, festivals, laws, dress, food, and housing. To these original existing practices, he says, should be added a conception of a rational social order, and "you have *li* in its most complete sense" (p. 225). Confucius considered *li* to be an indication of the moral strength of a nation. In its highest sense, it means "an ideal social order with everything in its place, and particularly a rationalized feudal order" (Lin, 1938, p. 13).

Knowledge is yet another concept in Confucius's philosophy. He defined knowledge as knowing people and as seeking to understand human nature in its context and in individual and social lives. Knowledge was regarded by Confucius as a way of self-perfection, of self-cultivation (Oliver, 1971, p. 132). "In its highest development, knowledge was to become wisdom, able to comprehend particulars through a unified insight" (Cleary, 1992, p. 6).

Such is but a very sketchy examination of Confucius's philosophy. Due to the seemingly unsystematic nature of his writings, it is hard to present a comprehensive picture of his philosophy, but Cleary (1991) has presented us with a good summarizing statement:

> The glue that binds everything together in the pragmatic moral universe of Confucius is the virtue of truthfulness or trustworthiness, faithfulness to the ideals exemplified by the sum of the cardinal virtues of humanity, justice, courtesy, and wisdom. Confucius likened trust to the link between a vehicle and its source of power and taught that trust was absolutely essential to the life of a nation. (p. 6)

Confucianism has influenced China for about 2,500 years. Ignoring its influence by any researchers of the history of Chinese thought and culture would be a grave mistake. Lin (1938) attributed three factors to the tremendous impact of Confucianism in Chinese history:

> first, the intrinsic appeal of Confucius's ideas to the Chinese way of thinking; second, the enormous historical learning and scholarship accumulated and practically monopolized by the Confucianists, in contrast to other schools which did not bother with historical learning (and this body of scholarship carried enough weight and prestige of its own); and thirdly, the evident charm of personality and prestige of the Master himself. (p. 24)

The magnitude of the influence of Confucianism is so much so that it has pervaded the speech and actions of almost every ordinary Chinese without him/her necessarily being aware of it.

Of all the aspects of Confucianism, the most relevant to my study is its rhetorical implications. There have been various, sometimes conflicting, studies on Confucian rhetoric. Haixia Wang (1993) provides a succinct summary of these studies. According to Wang, Confucian rhetoric is characterized by three distinctive traits: communal, historical, and dialogical. Wang defines communal as Confucian rhetoric that is based on communal understandings and interpretations of moral notions rather than a specific set of precise criteria. Non-deductive reasoning is favored as well as deductive reasoning. Confucianism emphasizes the concept of the community of like minds, that is, members of a community sharing common viewpoints on certain issues. Discourse that strays from such communal understandings is not likely to be effective rhetoric.

Confucian rhetoric is historical, according to Wang (1993), in that "each individual . . . is responsible for decisions regarding how exactly in each specific situation the principles of Confucian ethical notions will be applied" (p. 44). There is no set of self-consistent rules for moral judgment. The same is true with Confucian rhetorical principles. The rhetor is left to make his/her own judgment in response to exigent situations of discourse.

The communal and historical nature of Confucian rhetoric inevitably makes it also dialogic. The need to interpret communal understanding and to make judgments in exigent situations renders Confucian rhetoric a negotiation and argumentation among the Confucian rules about the meaning and the implementation of Confucian ethical notions in specific situations. "The necessity to discuss these implementations and the possibility that individual agents may be arbitrary make dissensus within a consensual community necessary, thus the inevitable dialogic nature of Confucian rhetoric" (Wang, 1993, p. 44).

I have devoted quite some length to the discussion of Confucianism because, as one of the most dominant ideologies in the history of China, its impact on people's perceptions of various social phenomena, including technology development, is readily felt in the development of major writing technologies in China, which I will show in the next chapters.

Taoism

Taoism was founded by Lao Tzu (also spelled as Lao Zi), who lived in the times of Spring and Autumn (770 BCE–476 BCE), and one of his students and his successor, Chuang Tzu (also spelt as Zhuang Zi). They are the two most profound thinkers in the history of Chinese philosophy who have been least understood by people. Yet, their philosophy has been as influential as Confucianism to the Chinese culture, if not more so. Some Western researchers have even traced a part of the Western philosophical tradition to Taoism.

To understand Taoism, we must first of all know what Tao (or Dao) is. Unfortunately, since Lao Tzu and Chuang Tzu never really defined Tao in clear-cut terms, any attempt to do so might do it an injustice. Nevertheless, this should not deter our efforts to understand its essence. Essentially, Tao refers to the "Way," the way of things. It makes possible what is impossible: "Tao gives birth to one, one to two, two to three, and three to everything else" (Lao Tzu, 1963, Ch. 42). It identifies itself with inferior and negative terms such as the weak, the low, etc. It is characterized by its identity as nonbeing (wu), no-name (wu-ming), no-form (wu-xing), not-having (wu-you), not-striving (wu-wei), not-knowing (wu-zhi), etc. It could be interpreted as the mysterious almighty creator, the basic element in the cosmos, an equivalent to the Western logos, the totality of all things or the ultimate reality, natural laws or the order of nature, the Great One, elusiveness, something undefined yet complete, chaos, or non-being. Or, none of the above.

Such an elusive nature renders Tao as something formless, inaudible, invisible, intangible, distant, and vague. Yet it is this very elusive nature that has intrigued people, that has drawn people to study or follow its principles. Viewed from a different perspective, this elusiveness can also be seen as flexibility, and it is this seemingly infinite flexibility that enables people to explain so many of the mysteries and the seemingly unexplainable phenomena in this world. Tao is the state of non-opposites that strike the balance between things; Tao is a transformational tool that nurtures interchangeability and, consequently, relativism. Tao aims at arriving at dialectic relations between dichotomies, yielding to and helping strengthen a rising power in order to hasten the moment of its decline; ultimately, it strives to achieve a whole and complete entity, an ultimate or essential realm, by means of a certain mystical or Taoist transcendence beyond the whole problem.

Taoism is engaged in the reversal or inversion of the metaphysical tradition. However, such a reversal or inversion is only the initial stage rather than the end of the deconstructive process. They do not aim at privileging the "other" term, but rather problematizing the reification, uncovering the interplay, and opening up the closure of binary oppositions in metaphysical thinking. In this sense, Taoism shares many principles with Derrida's theory of deconstruction, which is why it is all the more important to understand Taoist precepts, as they readily apply to many contemporary social phenomena. What is distinctive about Taoism is that it goes one step further than Derridean deconstruction in that it aims at unifying the differences so that one can ultimately be immersed into nature.

Taoism points out that once the world is differentiated in terms of dichotomies such as right/wrong, good/bad, life/death, beautiful/ugly, success/failure, gain/loss, more/less, long/short, big/small, etc., man's use of such terms will inevitably carry with it an immediate value judgment on experience, which then, to a great extent, affects his attitude and action. Taoism posits that all values are relative, perspective-bound, and arbitrary because they are not the intrinsic nature of things. The dualistic conceptualization prevents us from seeing the internal structure of a thing, which is to be understood in terms of differences. If from this internal point of view these dualistic terms become reversible and interchangeable, then there is no more opposition between them. The opposite terms become interdependent and complementary at a deeper level: without "that," there is no "this': without "this," "that" has nothing to hold onto. According to Taoism, terms are mutually defined. One does not have any value without the other:

> There is a beginning. There is a not-yet-beginning. There is a not-yet-beginning-to-be-a-not-yet-beginning to be a beginning. There is being. There is nonbeing. There is a not-yet-beginning to be nonbeing. There is a not-yet-beginning-to-be-a-not-yet-beginning to be nonbeing. Suddenly there is being and nonbeing. But between this being and nonbeing, I don't really know which is being and which is nonbeing. (Chuang Tzu, 1965 Ch. 2)

This famous quote is probably the best summary and representation of the Taoist perspective of relativism. What Chuang Tzu is saying here is that things, including those that are posited as opposites in dichotomies, are dependent upon one another to have an existence of any meaning or significance. Viewed this way, nothing is really more significant than the other, which then allows no necessity for the privileging of one thing over another. The following story from Chuang Tzu best illustrates the principle of relativity in Taoism:

> When Zhuangzi's [Chuang Tzu] wife died and Hui Shi came to convey his condolences, he found Zhuangzi squatting with his knees out, drumming on a pan and singing. "You lived with her, she raised your children, and you grew old together," Hui Shi said. "Not weeping when she died would have been bad enough. Aren't you going too far by drumming on a pan and singing?"
>
> "No," Zhuangzi said, "when she first died, how could I have escaped feeling the loss? Then I looked back to the beginning before she had life. Not only before she had life, but before she had form. Not only before she had form, but before she had vital energy. In this confused amorphous realm, something changed and vital energy appeared; when the vital energy was changed, form appeared; with changes in form, life began. Now there is another change bringing death. This is like the progression of the four seasons of spring and fall, winter and summer. Here she was lying down to sleep in a huge room, and I followed her, sobbing and wailing. When I realized my actions showed I hadn't understood destiny, I stopped." (Ebrey, 1993, p. 31)

Let me try to summarize Taoism without doing it too much injustice. Lao Tzu and Chuang Tzu's Tao can be understood as the way of life or the laws of nature. Tao ultimately leads to everything. Everything is made up of two opposites, which can transform into one another. The formation and transformation of everything represent the unity of being and non-being. Being and non-being depend on each other, but the non-being is more fundamental. Everything results from being, which, however, results from non-being.

Perhaps it makes more sense if we compare Taoism with Confucianism. While Confucianism emphasizes rational understanding,

Taoism depends on feeling and intuition. Confucianism sees truth and knowledge as being out there whereas Taoism regards the nature of truth as uncertain and conceived knowledge and truth as products more of perception. Confucianism values clarity in speaking and was against sophistries and vagueness; in contrast Taoism sees perceived vagueness as a virtue and considers argumentation to be futile.

In brief, while Confucianism represents the orthodox of the Chinese culture with its traditions and the master narratives, Taoism seems to reflect the more unorthodox traditions in the Chinese culture and to account in a better way for the unaccountable. What Taoism tries to do is to reconcile the orthodox with the unorthodox, the unaccountable, the marginal, and bring them to unity. The way of doing this is through the unity of our way of life with the laws of nature. What is significant about Taoism is its emphasis on the equal importance of both the orthodox and the unorthodox, thus granting space to the marginalized. This is what attracts people who subscribe to Taoism, and many people find Taoist principles especially applicable today when chaos and the unaccountable seem more than rare occurrences.

Rhetorically speaking, the Taoist view of language is characterized by a more radical "historicism or situatedness" (H. Wang, 1993, p. 54). Taoist use of language, argues H. Wang, "relies on human spontaneity, which is guided by human reasoning" (p. 54). This spontaneous nature of language use results from the fact that, while things in nature are constantly changing, man's limited perception of them reflects only the present:

> Since the linguistic systems are actually conceptual frameworks that organize and articulate our experiences with the world, to the extent that these frameworks spontaneously sort out only fluid boundaries and relations among momentary differences according to the circumstances, it is helpful; the effort to fix these boundaries as people often do in argumentation, however, is not helpful. (H. Wang, p. 57)

Understood this way, Taoism situates discourse in concrete social and historical contexts, and Taoist rhetoric is essentially "dynamic, creative, and individualistic" in nature (H. Wang, p. 58).

Buddhism

The main reason that Buddhism, one of the most dominant religions in the history of China, finds its way into my discussion of ideologies involved in technology development and transfer is that, although a religion, it is no doubt one of the most influential ideologies that has impacted the Chinese of all kinds, from the intellectual elite to the uneducated, from high government officials to laymen, from the wealthy to the poor. As Robert Somers (1990) has claimed, "few aspects of Chinese life, from high politics to popular culture, were untouched by the increased influence and mass appeal of this sophisticated religious system and the church that articulated and disseminated its teachings" (p. x). Buddhism "worked its way into all the domains of Chinese life—fundamental social doctrine, systems of belief, political institutions, and every sphere of culture, including architecture, sculpture, and painting" (Somers, p. x).

Buddhism first originated in India and later spread to China. Its exact date of introduction into China is an issue of dispute. It has been variously identified as around the first half of the first century CE (Fung, 1948), or the second half of the first century CE (Wright, 1990), or the second century CE (Xia et al., 1979), with some claims as specific as 67 CE (Pachow, 1980) and others as vague as the period from the first to the third centuries CE (Ebrey, 1993). Buddhism was commonly believed to have been founded by Shakyamuni Buddha (ca. 563–483 BCE). "As a set of ideas, it built on the Indian conviction that sentient beings transmigrate through endless series of lives as people, animals, gods, hungry ghosts, hell dwellers, or titans, moving up or down according to the karma, or good and bad deeds, that they accumulated," Ebrey (1993) explains. "The major insight of the Buddha was that life is inevitably unsatisfactory because beings become enmeshed in the web of their attachments. Yet he offered hope, teaching that it was possible to escape the cycle of rebirth by moral conduct, meditative discipline, and the development of wisdom" (p. 97).

After its introduction into China, Buddhism branched off into many different schools. In spite of their differences, they generally agree on the theory of Karma, translated as Ye in Chinese and deed and action in English. Such a translation, however, does it much injustice, as its actual meaning extends far beyond simple deed or action to cover one's speech and thoughts as well. Whatever one does, says, or thinks causes some effect, whether in the present or in the future, and "the being of an individual is made up of a chain of causes and effects" (Fung, 1948, p. 243). This chain of causes and effects spans a long cycle. "The present life of a sentient being is only one aspect in this whole process. Death is not the end of his being, but is only another aspect of the process. What an individual is in this life comes as a result of what he did in the past, and what he does in the present will determine what he will be in the future. Hence, what he does now will bear its fruits in a future life, and what he will do then will again bear its fruits in yet another future life, and so on *ad infinitum*. This chain of causation is what is called *Samsara,* the Wheel of Birth and Death. It is the main source from which come the sufferings of individual sentient beings" (Fung, 1948, p. 243–44). Only in the course of many rebirths can man accumulate *Karma* and attain an emancipation of the self, the *Nirvana*.

According to Buddhism, the transcendence of the cycle of life and death (i.e., becoming a Buddha) is considered one's highest, ultimate attainment. The path to this apex is one's cultivation of the mind, which depends on endless good deeds and five prohibitions: no killing, no robbery, no adultery, no lie, and no alcohol, the meanings of which are compared by Wei Shou, a Chinese historian of the sixth century, to the five Confucian virtues of benevolence, righteousness, propriety, wisdom, and trustworthiness (Ebrey, 1993). Such a comparison, which seems to put Buddhism and Confucianism on the same philosophical plane, is certainly open to debate as many scholars in Chinese philosophy will attest to the obvious differences and contentions between Buddhism and Confucianism (see, for example, Ch'en, 1973; Chan, 1985). Nevertheless, it points to the fact that Buddhism, one of the most dominant religions and prevailing philosophies in the history of China, has had its fair share of ideological influences in various spheres of Chinese life.

While the real meanings of Buddhist principles, in spite of my obviously simplistic attempt to summarize them, are far from being transparent, its impact on China is undeniably huge. Its impact is such that, as Arthur Wright (1990) has claimed, "an understanding of Buddhism in Chinese history helps to explain and clarify the whole of China's development, that without such an understanding much remains inexplicable, [. . .] that the observation of Buddhism in interaction with Chinese cultural elements serves to bring into bold relief those institutions, points of view, and habits of mind which are most intractably and intransigently Chinese" (p. 32–33). Of course, as Wright has also asserted, a thorough understanding of the impact of Buddhist principles on various facets of the Chinese life requires the study of the following aspects:

- the history of the relationship between Buddhism and Chinese law
- the history of the state policies and institutions for the control of Buddhism
- the history of the relation between Buddhism and religious Taoism
- the history of Buddhism in relation to Chinese philosophy
- all aspects of Buddhism in relation to the total culture of a specific period. (p. 33)

Lack of study in these areas, however, should not deter our effort directed at understanding Buddhism. As my brief summary of Buddhist principles has demonstrated, it is not impossible to catch at least some occasional glimpses, if not anything else, of some of the essences of Buddhism, which should shed much light on our discussion of the development of some of the writing technologies in China.

A discussion, however brief, of these three most dominant ideologies in the history of China begs the natural question: how do Confucianism, Taoism, and Buddhism coexist, and how do they fare with one another? While the coexistence of these three ideologies is a known fact, their coexistence has not exactly been one of harmony and accord. In the previous section, I discussed the differences between Confucianism and Taoism, so here I will mention some major arguments that outline the distinction and tension between Buddhism and the other two ideologies.

Since the Buddhist concept of Karma and retribution identifies the individual as the cause of all evil, Buddhism encountered strong resistance in medieval China. Buddhists then engaged themselves in strong defense. In her analysis of *Xiaodao Lun* (*Laughing at the Tao*), an anti-Taoist polemical text written by the official Zhen Luan and presented to Emperor Wu of the Northern Zhou Dynasty in 570 CE, Livia Kohn (1995) observes four distinct lines of reasoning in defense of Buddhism. The first was the argument for difference, which countered the notion of equality between Taoism, Confucianism, and Buddhism and strongly insisted that Buddhism was fundamentally different from the other two. The second was the Buddhist claim that its teaching was superior "because of its transcendent and otherworldly nature." The third, which seems to be in direct contradiction to the first two, was its claim that Buddhist teaching was useful in the Confucian state and was more original and effective than Taoism. The fourth argument went one step further to claim that the sages of ancient China were in fact disciples of Buddha and that Buddhism was therefore the very foundation of the Chinese intellectual and social scene instead of being merely one of its additions (p. 38–39).

Buddhist arguments that were specifically directed toward Taoism, claims Kohn, were even more aggressive. One argument accused Taoists of inconsistency between their preaching and practice, claiming that Taoists preached noble philosophies while practicing dishonesty and vulgarity in their ways of dealing with the world. A second Buddhist argument dismissed Taoism as politically useless and harmful, arguing that Taoism, with its betrayal and rebellious practices, was a potential source of political instability and moral degradation of social virtues and thus formed a destructive link in the fine equilibrium of social, natural, and political forces. A third argument by Buddhism "invoked Confucian morality and common sense to move against specific Taoist ideas and practices," charging that the Taoist practices, as in the interaction between yin and yang, the swapping of wives as part of their ritual, and the use of alcohol, all went against Confucian morality (Kohn, 1995, p. 40). A fourth argument was fired at Taoism, against its claim to universality, arguing that the concept of Tao as the source of all things and as the key to understanding the world was misleading and could endanger the wellness of the state and the empire.

Such were the representative arguments against Taoism, and sometimes against Confucianism, in defense of Buddhism. *Xiaodao Lun*, the above-mentioned anti-Taoist polemic text, uses all these positions and is, in a way, a conglomerate of all these self-defense Buddhist arguments. Challenging the presumption that Taoism is the best teaching for the Chinese culture, *Xiaodao Lun* points out the inappropriateness of Taoism for such a role due to its absurdity, nonsensicality, and inconsistency. This work is an illustrative example of the Buddhist self-defense against the threat from orthodox Confucianism and popular Taoism and, in a way, reflects the curious phenomenon of contentious coexistence between Confucianism, Taoism, and Buddhism throughout most of the Chinese history.

Due to the fact that Buddhism in China is represented by many different schools of thought, and interpretations of Buddhism have been various and sometimes conflicting, it is extremely difficult to summarize Buddhist rhetoric. Nevertheless, it is not impossible to capture a glance at the essence of the Buddhist view of language. Because Buddhists believe "things neither exist nor non-exist in language as 'nameable constituents,'" they argue that "the ultimate truth is beyond the capacity of language" (H. Wang, 1993, p. 62). "However, they have a firm belief in the crucial role and necessity of language in pursuing the truth in this world" (H. Wang, 1993 p. 65). The dual nature of things, which are at the same time both real and unreal, makes it necessary for us to communicate things and experiences, and language as a "skillful and expedient means to express the supreme truth" becomes a necessary tool, however inadequate this tool might be. The Buddhist view of language, points out Haixia Wang, therefore represents a balanced position between the deconstructionist and foundationalist extremes. Such a rhetorical perspective is in some way an integration of other traditional Chinese rhetorical perspectives, including the Confucian and Taoist rhetorics.

It should be noted, though, that, as I mentioned earlier, there have been other, relatively less influential ideologies, which, however, played important roles in their respective historical periods. For example, Maoism dominated the Chinese ideological scene for several decades in the contemporary history of China. Its influence certainly can in no way be discounted when we consider the recent development of writing technologies. Such a plethora of ideologies makes it almost a certainty that several ideologies might be at work during any given historical period. This, in turn, complicates the issue of technology transfer and development and, at the same time, makes it all the more important to consider the role of ideologies in this complicated process.

2 (Un)loading Technology

> There is an enduring problem in our understanding of modern technology and its social significance. On the one hand, we have abundant empirical knowledge about the history of technical innovations, including the conditions of their development and their diffusion within and among different nations. On the other hand (the speculative side of the matter), the evaluation of the social consequences of technological change remains trapped in the seemingly arbitrary polarization between subjective feelings of pessimism or optimism, between warnings of doom and complacent advocacy of the "technological fix." ... Both views are fatalistic, in that they regard social institutions as being forced to adjust to changes brought about by technological innovations, and both consistently ignore the reciprocal influence that conflicting social interests exert on the process of innovation and application. Neither of these formulations permit us to reconcile the widely divergent claims about the social significance of modern technology. (Leiss, 1977, p. 115–116)

Indeed, the two polarized views William Leiss describes here still dominate popular perceptions of technological change and the interrelationship between technology and society. Though less trapped in a Cartesian type of polarization, researchers of technological development have barely traversed beyond the struggle to find a balance between the two extremes. While we have come a long way in our understanding of technology in the last five or six decades, we still find ourselves more or less in the same situation of being faced with the challenges to define technology with confidence and certainty, to delineate the connections between technology and culture, and to operationalize our theoretical perspectives so as to better guide technological development. The only difference is that this is an even more uncomfortable situation, as better knowledge about technology on our part has unearthed more dimensions and complexities about technology.

TECHNOLOGY LOADED

The term *technology* has a deceiving nature: on the one hand, the popular perception sets alarmingly simple parameters for its denotations: basically any set of human artifacts that extend the capabilities of man; on the other hand, an exploration into such parameters easily reveals high complexities associated with its connotation. A historical, as well as cultural, deconstruction of the term, therefore, is in order.

Originating in the seventeenth century, *technology* is used "to describe a systematic study of the arts or the terminology of a particular art" (R.Williams, 1983, p. 315). A distinction, however, should be made between *technology* and *technique*. According to Raymond Williams (1983), *technique* is defined as "a particular construction or method" whereas *technology* refers to "a system of such means and methods" (p. 315). William Leiss (1977) makes a similar distinction. What become technologies, according to Leiss, are "those transmissive techniques that attain a certain level of general significance in particular societies or historical epochs" (p. 119). The main reason for this distinction, he argues, is "that only the general modes of social organization, and not the specific properties of techniques themselves, determine which types of techniques will be encouraged and promoted and which will be downplayed or perhaps forbidden" (p. 119). What is important about this distinction is that *technology* defined as a system implies a greater complexity—the involvement of social organization—than what a single *technique* entails. It is this complexity that we must first figure out.

Over the three centuries since its incorporation into the language, the term *technology* has evolved through basically three different conceptualizations. A first, simplistic definition of *technology* sees it as mere physical artifacts. A second, more sophisticated version views it as including not only the physical objects but also the technical know-how. A third perspective considers technology a sociotechnological phenomenon and takes into account the cultural, social, and psychological factors involved in the development and use of technology. Langdon Winner's (1977) definition of technology offers a good summary of the distinctions between the different conceptions of *technology*. As Winner sees it, there are three different aspects of technology. The first is "tools, instruments, machines, appliances, weapons, gadgets"; the second refers to "the whole body of technical activities—skills, methods, procedures, routines—that people engage in to accomplish tasks and includes such activities under the rubric *technique*"; and the third involves "some (but not all) varieties of social organization—factories, workshops, bureaucracies, armies, research and development teams, and the like" (p. 11–12)

An understanding of technology, according to Frederic Fleron (1977), must start with the assumption that technology is "a sociotechnical phenomenon and an integral part of the basic values of every culture" (p. 12). Understood this way, it is not hard to see that "technological change in any culture has direct consequences for other cultural values" (p. 12).

This cultural aspect of technology is clearly defined in Leiss's (1977) argument. According to Leiss, technologies, as combinations of techniques, "represent choices among alternative uses or goals," and the choices as to which technologies to use and which to suppress are made based on "particular features of the institutional forms that predominate in a given society at a given time" (p. 120). Technologies are almost always a combination of techniques with class, status, and role determinations that dictate the uses of technologies and the forms of their use. It is this combination, argues Leiss, that defines technology, that renders technology concrete by giving it an operative context, and that gives technology a social character (p. 120–121).

A few researchers have attempted to define this social/cultural aspect of technology. Fleron (1977) describes it as "technical rationality," which, combined with "machines and tools themselves," forms technology (p. 3). He defines "technical rationality" as "that mode of purposive rational action that accompanies the implementation and use of a particular machine technology" (p. 3). Leiss (1972) defines it as "the purposeful organization and combination of productive techniques, directed either by public or private authorities" (p. 199). Both definitions point to aspects of social and production relations necessarily present in any given operative context of technology.

Andrew Feenberg's (1991) *Critical Theory of Technology* is even more revealing in this regard, in that it defines technological rationality as political rationality:

> The values and interests of ruling classes and elites are installed in the very design of rational procedures and machines, even before these are assigned a goal. The dominant form of technological rationality is neither an ideology (an essentially discursive expression of class self-interest) nor is it a neutral requirement determined by the "nature" of technique. Rather it stands at the intersection between ideology and technique where the two come together to control human beings and resources in conformity with what I will call "technical codes." Critical theory shows how these codes invisibly sediment values and interests in rules and procedures, devices and artifacts that routinize the pursuit of power and advantage by a dominant hegemony. (p. 14)

Thus, according to Feenberg, technology is an "ambivalent process of development suspended between different possibilities" (p. 14). Social values play a role not only in the use of technical systems, but also in their design. Technology in this sense goes far beyond its physical being; it becomes a "scene of struggle," "a social battlefield," or even "a *parliament of things* on which civilizational alternatives are debated and decided" (p. 14). In a word, technology is a site of rich social activities; it is a phenomenon loaded with social and cultural complexities.

Technology Unloaded

Recognizing that technology is loaded with complexities is only our first step toward a more critical understanding of the rhetorical underpinnings of technological developments. For the logical next step, and for the purpose of our discussion, let's do an artificial fragmentation of technology (i.e., unloading the term) but only temporarily.

Ever since recognizing that technology entails more than mere physical artifacts, researchers have made efforts to define the intangible aspect of technology in an attempt to arrive at a better understanding. Seeing both the tangible and the intangible aspects of technology,

Fleron (1977) labels the intangible component "technical rationality" and defines "technical rationality" within the capitalist context as:

> a broad set of basic cultural values including the dichotomization of means and ends, of work activity and product; the acceptance of means-ends efficiency as a primary goal; the compartmentalization of knowledge; the separation of mental and manual labor; the human domination of nature; efficiency defined in nonhuman terms; the hierarchical control of production; and underlying it all, the assumption of the priority and insatiability of human material wants. (p. 3–4)

Essentially, this capitalist technical rationality, in Fleron's conception, concerns primarily the capitalist cultural values about production relations.

When applied in a more general social context, technical rationality, according to Leiss (1977), is "the purposeful organization and combination of productive techniques, directed either by public or private authorities" (p. 199). In this definition, Leiss likewise points to something within technology that is other than the technical artifact itself—the cultural infrastructure, out of which the technical artifact is cultivated. This cultural infrastructure is an inherent part of technology and has a shaping impact on the form and function of the technical artifact.

A more illuminating approach is to examine this issue from a slightly different, yet integrally related, perspective—the technology transfer process. Any technology, from the design stage to the use phase, undergoes a transfer process, either from the lab to the marketplace or from one culture to another. A distinction has to be made, though, between the narrow and broad senses of the term currently being used in research in this field. In the (relatively) narrower sense, *technology transfer* refers to the transfer of a technology from one culture to another, from one society to another, and often from one country to another. This is what is intended by such researchers as Frederic Fleron (1977), Andrew Feenberg (1991), William Leiss (1977), Richard Baum (1977), R. F. Dernberger (1977), Rensselaer Lee (1977), William Dunn (1977), and Frederik Williams and David Gibson (1990), whose studies focus mainly on technology transfers from Western cultures to Eastern cultures. In a broader sense, *technology transfer* also refers to the transfer of technology to the marketplace as well as that from one culture to another. This is the sense employed by Stephen Doheny-Farina in his 1992 study *Rhetoric, innovation, technology: Case studies of technical communication in technology transfers*. It is also implied, though not explicitly stated, in other people's studies (e.g., Yates, 1989). This broader sense of technology transfer basically covers all transfers of technology from one context to another.

Looking at technology in terms of its transfer process provides a unique perspective in deconstructing the cultural aspect of technology. In this respect, Williams and Gibson's (1990) interpretation of what technology transfer implies provides a revealing example. Defining technology transfer as "the application of knowledge" and therefore an act of communication (p. 10), Williams and Gibson articulate more clearly the complexities involved in this communication process:

> The communication involved in the technology transfer process often takes place between individuals using different vocabularies, styles, channels, schedules, and reward systems. Formal and informal communication barriers exist between different cultures, different sciences, and different levels of abstraction in what is being transferred (e.g., an abstract theoretical technology as against, say, a concrete product design). Serendipity may play a larger role in effective transfer than most people would like to admit; an important link may arise from a friendship, a crisis consultancy, or office or laboratory layout. Technology transfer does not necessarily lend itself to a rational mathematical model. (p. 13)

The communication barriers described by Williams and Gibson have often times resulted in transformation of the technology such that the transferred technology becomes something different in the target culture from what it was in the originating culture. In most cases, however, the technical artifact itself is unlikely to have undergone metamorphosis to any serious degree (though in some cases it would be modified to fit particular purposes in the new culture). If this is true, then the differences must have come from something within the technology that is other than the technical artifact itself.

Therefore, it is reasonable to assume that such communication barriers and the consequent change in technology are a result of the differences between the source and target cultures, be they local (such as corporate cultures) or global (ethnic and national cultures).

Take the computer for example. Though a keyboard on a computer used in China might look exactly the same as that used in the U.S., its functions might be dramatically different in that the same English letters are used in combinations to produce Chinese characters. The keyboard in China is so configured obviously because of the nature of the Chinese language—each character requires two bytes of space in the ASCII system, whereas each English letter takes only one. Thus, the keyboard functions very differently in these two different cultural contexts, although its appearance remains more or less the same.

Therefore, simplistically speaking, technology transfer has been argued to be the moving of a particular piece of technology from one culture to another as illustrated in Figure 1. In this linear view, technology transfer seems to be simply migrating from one location to another, without any transformation whatsoever.

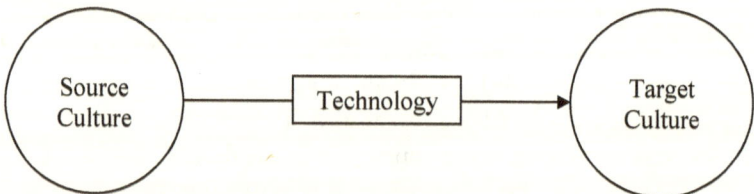

Figure 1. Technology Transfer—A Linear Perspective

If this transferred technology represented an identifiable and distinguishable unit only, we would be able to easily define its intrinsic elements and isolate them from the other elements in the society. However, as we mentioned earlier, there is a second aspect of technology that involves more than the technical artifact itself—the complex infrastructure of the culture. Thus, we get a slightly more complicated view of this transfer process (see Figure 2).

This cultural aspect of technology—technical rationality—does not lend itself to easy definition and simple interpretation. If technology were a simple combination of technical artifact and technical rationality, juxtaposed within the technology "container," our issue here would be far less complicated. The problem is, instead of being

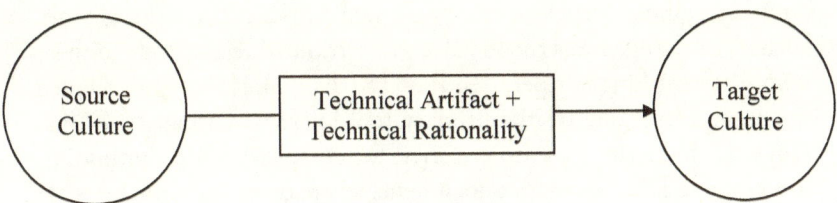

Figure 2. Technology Transfer—A More Complex Linear Perspective

two distinguishable, separate identities confined within a larger identity, the two are meshed together with no visible boundaries between them. Therefore, technology as it operates in a culture is not a simple juxtaposition of technical artifact and technical rationality. Nor is it a simple overlapping of the two elements, even though they can overlap to different degrees. Instead, it is, more precisely, an immersion of the two within and an infusion of one another. As technical rationality represents the cultural elements of technology, the marriage between technology and culture is thus inevitable. Since technology is culturally signified, as Andrew Feenberg (1977) puts it, as a means for the "actualization of human capacities," "it will be pursued spontaneously by the individuals as a positive component of their own welfare and need not be imposed on them by market incentives or political or moral coercion in opposition to their own perceived interests" (p. 103). Consequently, argues Feenberg, technology is made to conform to the changing economic goals and the needs of a culture and thus "is not destined to remain a fixed and antagonistic element never fully assimilated by the culture into which it is inserted" (p. 103). This bonded relationship between technology and culture is more clearly defined in Feenberg's 1991 work, *Critical Theory of Technology*, where Feenberg defines the technology transfer process basically as a process of first decontextualization and then recontextualization. The complexity is that this bonding happens not at a particular, specified moment, but through a gradual process during which elements on both sides fuse into one another.

A revealing example is the different degrees of significance attached to the use of fountain pens versus ballpoint pens in the U.S. versus in China. Both were Western inventions and were imported to China during different historical periods, with the fountain pen considerably earlier than the ballpoint pen. In the transfer processes, what was retained about both kinds of pens was their function as writing tools, and they quickly replaced the brush pen as the main writing tools. What got changed was the social status attributed respectively to these two different kinds of pens. In the U.S., ballpoint pens are accepted for legal forms and are the primary tool for handwriting. In China, however, fountain pens are preferred for completing and signing important and legal forms and documents whereas the ballpoint pens are often considered informal and thus inappropriate for formal and legal occasions. Although this may have been due to the inferior quality of earlier forms of ink for ballpoint pens, a more subtle reason for the different status of fountain versus ballpoint pens may lie in the fact that calligraphy, a highly respected form of handwriting, has traditionally been associated with fountain pens (other than brush pens), but never ballpoint pens. This high status attributed to calligraphy may have been transferred to other situations of writing with the use of fountain pens. Thus, both kinds of pens, though the same in physical appearance and function, have quite different status in the two cultures.

To sum up, the study of technology transfer and development involves not only the technical artifact being transferred but also the technical rationality that comes with it as an inalienable element of technology. Loosely defined as the cultural infrastructure, technical rationality is a combination of elements from both the source and target cultures and is a product of the constant negotiation between the two. A study of technology transfer and development thus entails a thorough understanding of what particular cultural elements are at play in the process and how and to what extent they shape the development and meaning of technology.

What adds to the complexities of the whole issue is that during the process of technology transfer, the acquisition by technology of the elements of the new (target) culture is almost always accompanied by a loss of some of the elements of the source culture. What is more complicated is that this loss of the original cultural elements is seldom complete but almost always partial. Thus, the transferred technology retains part of the original culture while absorbing the new one. The two cultures within technology will likely clash at first encounter until they find some common ground on which to coexist and even merge into one another through a constant process of negotiation affected by the various factors within the context of technology transfer.

Why, then, is culture so important to technology?

So far, I have not really defined what culture is, and my implied definition of the term in relation to technology transfer and development—that culture here refers to the contextual, non-physical aspects of technology—is obviously limited. Therefore, before I answer the question about the relationship between culture and technology, let me first attempt to offer a broader, and hopefully more thorough, definition of culture, which I believe will better illustrate the complex nature of technology and its integral relation to culture.

Culture is probably one of the most difficult terms to define due to its loose and fluid nature. Williams (1983) traces the history of the development of the concept of "culture" to the early fifteenth century. From its early denotation to the tending of natural growth to the later sense of the process of human development, the term *culture* has come to encompass both the cultures of different nations and periods and the cultures of different groups within a nation (p. 87–89). According to Williams, culture, as it stands now, refers to the following three things:

1. a general process of intellectual, spiritual, and aesthetic development,
2. a particular way of life, and
3. the works and practices of intellectual and especially artistic activity. (p. 90)

Williams argues that this complex list of meanings indicates a complex argument about the relationships between the three, which in turn points to the complex nature of culture. What is obvious is the intangible nature of culture. As this three-item list indicates, at least two of these three things are non-physical, invisible, and intangible, and only less than one out of these three is visible and tangible. Understanding culture thus often means to infer its meaning based on some visible and tangible artifacts that constitute only a small part of its content.

Even such a small material, physical existence does not always have its place in the definition of culture, at least not in Richard Weaver's conception. According to Foss, Foss & Trapp (1985), Weaver's conception of culture does not include material goods but is of the imagination, the spirit, and inward tendencies. In Weaver's eyes, culture plays a role of control in our society. He sees a center of authority at the heart of every culture, which he calls the "tyrannizing image." This center represents a high ideal of excellence and perfection that a society strives for. Being a focus of value, a law of relationships, and an inspiring vision, it sets up a hierarchy of rankings and orders and a set of terms and codes that everybody is expected to follow. Thus culture exerts subtle and pervasive pressure upon its members to conform to its codes. It follows then that culture exerts control over people's actions. Nevertheless, Weaver does see some freedom for people to act under this cultural system, and he points out that more often than not people fall short of the ideal advocated by the culture. Thus, the final product that we call culture is always a result of negotiation and compromise between the society's code of standards and people's actual conduct.

Thus, what culture is becomes less stable than we might want it to be. This unstable nature, according to John Van Maanen (1988), is what characterizes culture. Van Maanen describes culture as a "loose," "slippery," and evolving concept. Although culture is generally defined as the knowledge of the members of a given group that shapes and accounts for the activities of the members of the culture, this concept is not static, tangible, or visible by itself. Culture, in much current ethnographic research, is described as "contested, emergent, and ambiguous" (Van Maanen, p. 127). It can only be understood through the actions and words of its members. Due to this ambivalent nature, culture is not something to be known once and for all: "knowing a culture, even our own, is a never-ending story" (Van Maanen, p. 119). Conceptualizations of culture have thus evolved from static conceptualizations to more fluid representations.

Complex and discouraging as it may seem, this fluidity of culture points less to its incomprehensibility than to its dynamic nature. What it suggests is that if technology embodies cultural values, then it is as much a dynamic entity as culture is.

Such complexities concerning culture and technology transfer point to two problems in current research on the development and transfer of writing technologies, which this study will address at the macro level. First, while there is much research on the impact of writing technologies on our writing behavior and culture, and while a recognition of the role culture plays in the development of writing technologies is evidenced in some studies (e.g., Doheny-Farina 1992; Dunn 1977; Feenberg 1977, 1991; Fleron 1977; Hoffman 1977; Williams & Gibson 1990), little has been done in understanding the nature and the magnitude of this role, the elements that facilitate this role, and the way of facilitation. Second, existing studies on technology development and transfer are based almost exclusively on the assumption of a one-directional flow from developed to developing countries. Ignored in such an assumption are (1) the possibility of a reverse transfer path and (2) the important issue of the native development of technology in developing countries. This possibility of a reversed technology transfer, however, is not the subject of this study. What I will focus on is the more important implication of such a possibility (i.e., the role of the native culture in domesticating the foreign technology). The purpose of examining this role is to explore, and hopefully to answer, the questions of what has accounted for the different paths of technology development in different cultures and what determining factors shape the construction of the meaning of technology.

Some researchers have long recognized the fact that culture plays a certain role in the development of technology, though their understanding of the nature and the magnitude of this role is to varying degrees. Leiss (1977), for example, recognizes "the reciprocal influence that conflicting social interests exert on the process of innovation and application" (p. 115–116). Fleron (1977) argues that the starting point for an understanding of technology is how cultural values are reflected in a particular technology, especially in its formative stages. JoAnne Yates (1989) focused her exploration of the development of communication technologies in nineteenth-century America on the cultural and social factors that contributed to the emergence of those communication technologies.

The role culture plays in the development of writing technologies is also implied in Paul LeBlanc's (1993) study of the development of several writing software programs including *Prose* and *Daedalus*. *Prose*, "a program designed to facilitate instructor feedback on student texts," was developed by Nancy Kaplan and her colleagues (LeBlanc, p. 34). *Daedalus*, a program that facilitates synchronous exchange on the computer, was developed by Fred Kemp and several other graduate students at University of Texas in Austin in the late 1980's. These two programs went through quite different development paths due to the different cultural contexts within which they were developed. One of the major contextual factors, for example, was corporate involvement, or the lack thereof, in the development process, which determined to a large extent the consequent fate of the programs. *Prose* "was sold by the designers to McGraw-Hill, effectively removing from their hands the future of the program" whereas "the Daedalus Group retains control over the fate of its programs and continues to develop and market them on its own". (LeBlanc, p. 51). Of course, the different fates of these two programs were the result of various factors in addition to corporate involvement, but the role of cultural context is more than obvious in these two cases.

Such an understanding of the role culture plays in the development of technologies, including writing technologies, is more clearly articulated in Andrew Feenberg's and Stephen Doheny-Farina's works. Feenberg's (1991) *Critical Theory of Technology* examines in depth the reciprocal relationship between culture and technology. He argues that social as well as technical criteria of progress shape the direction of technological development as much as technology shapes certain aspects of the culture. Culture, according to Feenberg, impacts technology in the contextualization of technology.

Doheny-Farina (1992) examines the role of culture in technology development from a different, and, for the purpose of my study, more illuminating, perspective—the rhetorical perspective. He looks at technology transfer and development as a rhetorical phenomenon. Technology transfer and development, according Doheny-Farina, are essentially a rhetorical process of knowledge construction, of conceptualizing, reconceptualizing, negotiating, and communicating meanings about technology. Such a perspective attributes a (well-deserved) prominent role to the cultural context in writing technology development. Applying such a rhetorical perspective of knowledge construction to the examination of technology development will explain the extent to which cultural factors shape and determine the meaning construction and development paths of writing technologies.

The significance of this research lies in its recognition of the role of culture in technology development and of the reciprocal relationship between technology and culture. In addition, the work of some researchers, especially Feenberg and Doheny-Farina, exhibits great depth in its examination of the nature and magnitude of such a role and relationship, and represents ground-breaking efforts in research in this area.

However, while this body of research has opened our eyes and showed us what the big deal is about culture, it has at the same time raised many issues. That is, while it has provided us with the understanding that technology transfer and development are essentially a cultural and rhetorical phenomenon, it has prompted us to ask many important questions. What cultural and rhetorical factors are at play in the transfer and development process of writing technologies? In other words, if we are to examine a particular technology transfer and development process, what elements exactly should we look at in order to understand what has constructed the meaning of that technology? What kind of model of analysis can and should we employ to understand that process? How can we operationalize the study of any writing technology?

Technology as Emergence

Understanding the role of culture in technology development has been a consistent theme in our efforts toward a systematic conceptualization of technological phenomena in the last few decades. Answering all the above questions, however, necessitates an examination of this conceptualization path we have traveled thus far. An appropriate place to start this examination is with the various theories of technology developed in the past half century or so, and this examination reveals several distinct patterns.

Before I elaborate on what these patterns are, let me borrow Steven Johnson's concept of emergence in defining technology development. Emergence, according to Johnson (2001), is the "movement from low-level rules to higher-level sophistication" (p. 18). Emergent systems are "complex adaptive systems that display emergent behavior. In these systems, agents residing on one scale start producing behavior that lies one scale above them [. . .] " (p. 18). The beginning of emergence is marked by "a higher-level pattern arising out of parallel complex interactions between local agents" (p. 19). This concept is also reflected in the work of Roger Levin (1992), who defines emergence in strikingly similar terms: emergence is "order arising out of a complex dynamical system, [. . .] global properties flowing from aggregate behavior of individuals" (p. 13). The way emergent systems work, explains Levin, is that the interaction of the individual components at the local level leads to the emergence of some kind of global property at the higher level, which in turn "feeds back to influence the behavior of the individuals [at the local level] that produced it" (p. 12–13). By Johnson's and Levin's accounts, any kind of systematic development can be defined as emergence, and technology development is certainly no exception, for several obvious reasons. For one, technologies are "complex adaptive systems that display emergent behavior"; for another, any technological development is a "movement from low-level rules to higher-level sophistication"; for a third reason, the interactions in the dynamic system of technology will eventually lead to a higher-level pattern, an order, of global properties.

Besides informing our conceptualization of technology, casting technology in the light of emergence also yields a more expedient result for the purpose of my discussion here: it illuminates on the patterns/phases of this conceptualization process. The process of emergence, argues Johnson (2001), is characterized by four phases/principles: (1) neighbor interaction, (2) pattern recognition, (3) feedback, and (4) indirect control (p. 22). "In the first phase, inquiring minds struggled to understand the forces of self organization without realizing what they were up against;" "in the second, certain sectors of the scientific community began to see self organization as a problem[...] partially by comparing behavior in one area to behavior in another;" "in the third phase [...] we stopped analyzing emergence and started creating it;" and in the fourth phase, we make attempts to control it" (Johnson, p. 20–21).

This process perspective will be applied here in two ways. First, synchronically, each particular technology that eventually matures inevitably goes through such a phased process as the human agent makes efforts to understand and interact with the technology. How this process evolves in the case of each technology will be demonstrated in my discussion of specific technologies in subsequent chapters.

Second, diachronically, the human perception of and relationship with technologies have undergone a similar evolving process in the last five or six decades. Starting in the mid-twentieth century, our efforts of systematic conceptualization of technology and of the human relationship with it have gone through four different phases: 1) passive exposure, 2) initial understanding, 3) reactive interaction, and 4) proactive control, a process much in congruence with Johnson's categorization. The mid-twentieth century is set as the starting point not entirely arbitrarily because no well-formulated theory of technology had existed until then. Although it's impossible to pinpoint the exact starting point of systematic theorization of technology development, the past half a century or so has witnessed the emergence of the following important theoretical conceptualizations of technology: the neutrality theory, the theory of technological determinism, the substantive theory, and the convergence theory, the theory of scientific and technological revolution, the ambivalence theory, the mediation theory, the critical theory, the social theories, and the rhetorical theories. This procession of theory development, which is far from linear, reflects the evolutionary, though sometimes recursive, changes in our understanding of, attitude toward, and interaction with technology, which can be categorized into the four phases mentioned above. An examination of these phases of development below serves to contextualize my theoretical framework and foreground the need to develop operationalized strategies to better cope with technological developments.

Phase 1. Passive Exposure

Obviously, technologies of various kinds have been a part of human life for thousands of years. However, little research of any serious nature had been conducted until about half a century ago. When philosophical research of any documented magnitude in technology took form in the mid-twentieth century, the human reaction to technological change was one of passive exposure. As a natural reaction to burgeoning technologies surrounding mankind, the neutrality theory of technology represented an easy way out of a complicated phenomenon when human cognition of the technological phenomenon was still negligible. Technology, according to advocates of the theory of technological neutrality, has "no particular consequences for social organization and the nontechnical aspects of culture" (Fleron, 1977a, p. 457). Several distinguishing features of the neutrality theory suggest that mankind was more or less being passively exposed to technologies without a serious understanding of their nature and impact.

First, the neutrality theory claims that there is no inherent ethical value to the nature of technology. No technology is born "good" or "bad" by nature. Technology is technology: it is neutral. Technology becomes "good" or "bad" only when it is used by man for good or evil purposes. The use of technology for a "good" or "evil" purpose in a particular case depends not on the nature of technology itself, but on man's manipulation of it. The technology stands aloof from its applications.

Second, along a similar line, the neutrality theory claims that technology has no inherent cultural value. Technology cannot be categorized as "capitalist" or "socialist." This claim essentially denies any non-technical aspect that could possibly be associated with technology. Technique, with its various elements understood as "technical means," "as opposed to the social consequences of its use, is regarded as neutral with respect to social systems and class," thus rejecting such notions as "capitalist technique" or "proletarian technique" (Cooper, 1977, p. 147). Technology carries no cultural identity and therefore no cultural baggage. Technology as a product is not affected by the process of its creation, by what in fact is a value-laden process of socially constructing meaning and knowledge about technology. According to such a view, technology transfer becomes a simplified act of transferring concrete techniques and/or physical machines which does not involve any cultural transformation or reconstruction of meaning.

Third, as an extension to the above two claims, the neutrality theory regards the use of technology as external to technology. The uses to which technology is put is not predetermined by any internal mechanisms or dynamics of technology because the lack of any internal ethical and cultural values does not allow technology to dictate by any means and to any extent the purposes of its application. Such a view is, as Fleron (1977a) points out, "eminently compatible with, and in fact historically and intellectually related to, the so-called fact-value distinction, by which facts and the material world are seen to stand 'out there' in full view and all readiness for scientific scrutiny and material progress, whereas values and political choices lurk in the shadows of human emotionality and/or the cloudy rarified regions of spirits and gods" (p. 458). This fact-value distinction, as Fleron notes, is derived from the many false dichotomies that characterize the modern Western philosophical tradition.

Simply put, what theorists mean by technological neutrality is, in Feenberg's (1977) terms, "that the technology developed under capitalism is simply neutral, that the same means can be used for different ends" (p. 108–109). It essentially dismisses the possibility of any intrinsic relationship between technology and culture. Such a perspective echoes the traditional approach that views all technologies as simply tools.

One flaw in the dismissal of technological impact on the society and culture is obvious. What has accounted for the neutrality perspective on technology is both its isolation and confusion between technology as a product and technology as a process. The neutrality theory sees technology as only a product. Excluded from this perspective is the aspect of technology as a process of development and use. The importance of viewing technology as a process as well as a product lies in that such a view recognizes "the social consequences of technological process itself" (Fleron, 1977, p. 459). To ignore the process aspect is to ignore the process by which technology shapes people and the culture in the same way that it is shaped by the people who create it, their values and beliefs, and the cultural environment that shapes the creators of technology in the first place. The relationship is one of mutuality and interactivity.

Of greater consequence, is this dismissal of technological impact by the neutrality theory when it is applied in the case of technology transfer. By denying the process by which cultural cultivation occurs, the theory of neutrality creates a false impression that technology does not carry with it any cultural marks and thus in its transfer from one culture to another is not likely to impact the target culture with its original cultural traits. Technology transfer examined in this light becomes a deceitfully oversimplified process. Simplistic as it is, such a position, that argues for little or zero impact of technology on society and culture, is still endorsed by some scholars today.

Phase 2. Cognitive Understanding

If neutrality theory was more or less an involuntary, hasty reaction as a result of our initial exposure to technology, the subsequent set of theories represented a more organized attempt at a deeper, cognitive understanding of technological phenomena. This set of theories includes the theory of technological determinism, the substantive theory, the convergence theory, and the theory of scientific and technological revolution. All of these theories assume, to varying degrees, that technological changes effect certain inevitable consequences and follow particular development paths that defy intervention from social and cultural forces.

Compared with the neutrality theory, the theory of technological determinism represents the other extreme on the continuum of technology theories. According to Fleron (1977a), the theory of technological determinism argues that "there is some form of inner logic or dynamic in modern industrial technology that results in similar social consequences regardless of the setting" (p. 459). The underlying assumption of this theory, Feenberg (1977) points out, is that "culture plays no significant role in shaping the history of technological development but can only motivate or obstruct progress along a fixed and unilinear track" (p. 103). This theory appealed to many researchers around the mid-twentieth century, probably because it provided an expedient rationalization and convenient excuse for man's passive acceptance of, and his seeming inability to cope with, the overwhelming changes brought about by technology (see Baum, 1977; Dernberger, 1977; Ellul, 1980; Field, 1977; Galbraith, 1971; Heidegger, 1977; and Lee, 1977). In his 1971 work, *The New Industrial State,* John Galbraith argues that in the face of technological development, man's choice is so limited that he can only decide what level of industrialization he wants, that once he opts for industrialization, technology operates to a similar result on all societies, and that the result of industrialization is inevitable and basically the same (p. 380). Simply put, the patterns of technological development in all societies and cultures are more or less predetermined and the same, so social and cultural settings play no significant role in its development path.

This notion of determinism, that technology develops along fixed paths regardless of social and cultural settings, is derived from the conviction by theorists of technological determinism that there are certain elements about technology that are inherent and constant and that never really change. Galbraith (1971) identifies them as the following six imperatives:

- Increasing span of time
- Increase in capital commitment
- Increasing inflexibility of commitment
- Specialized manpower
- Specialized organization
- Planning

These imperatives operate in a similar pattern. They are the inherent elements of technology that any target society has to digest in the process of technology transfer.

The substantive theory and the convergence theory argue along a similar vein and are little more than variant names for the same theory. For example, in his examination of the impact of modern industrial technology upon the organization, management, and social relations of Chinese industrial enterprises, Richard Baum (1977) claims that his study of the Chinese case more or less confirms the following three propositions of the convergence theory:

1. Technological development is essentially unilinear, that is, that there exists a relatively fixed path or sequence of technological innovation over which all industrializing societies must travel in their quest for modernity.
2. Modern industrial technology tends to impose similar organizational constraints upon all societies, regardless of differences in culture, ideology, or political institutions.
3. In the process of adapting to these universal constraints, human values and behavior patterns tend to converge in a technobureaucratic ethos of "amoral instrumentalism." (p. 315)

Countering claims about China's possibility and ability to bend the technological imperatives with the Maoist anti-bureaucratic, anti-elitist ideology, Baum claims that his survey of technical, managerial, and social structures and relations in Chinese industrial enterprises yields little evidence that China has succeeded in doing so. On the contrary, it suggests "that industrial technology, organization, management and human relations in the PRC are currently evolving in directions that are neither so fundamentally unique and innovative nor so radically egalitarian as China's more ardent apologists have asserted" (p. 351–352).

The problem with the theory of technological determinism, the substantive theory, and the convergence theory is that these theories embrace an ahistorical, acontextual perspective on technology development. Such a perspective assumes a one-way effect in the interactive relationship between culture and technology, denying any possible role that culture can play in technology development. It is thus an essentially rigid and pessimistic view of technology because, as Feenberg (1977) points out, it asserts a fixed pattern of technical progress for all societies and cultures and it "asserts that social organization must adapt to technical progress at each stage of development, according to 'imperative' requirements of technology" (p. 72–73). Such a perspective could have serious negative consequences when applied in real situations of technology transfer and development, because it denies man, the most important element in the development and transfer process, the due role of rhetorically constructing and reconstructing the meanings of technology. Comparatively speaking, it can be an even more oversimplified and misleading perspective than the neutrality theory.

A more unique theory, the theory of scientific and technological revolution (STR), developed mainly by Soviet theorists, is worthy of note because of its applicability to socialist cultures. As a serious attempt to understand the relationship between science and technique and the rapidly changing modern world, the STR theory originated in the early twentieth century and became a widely accepted and practiced concept in the Soviet Union in the 1950s, 1960s, and 1970s. Julian Cooper's 1977 article, "The Scientific and Technical Revolution in Soviet Theory," provides a good account of the evolution of this theory in the former Soviet Union. According to Cooper, the essence of the STR is best summarized in the following definition provided by members of the Academy of Sciences Institute of the History of the Natural Sciences and Technique in Moscow:

> The replacement of the direct production functions of man, including his logical and control-regulation functions, by technical means is the essence of the present-day STR, as a result of which the technical conditions for the transition from machine-factory production to comprehensively automated production appear. (cited in Cooper, 1977, p. 156–157)

The theoretical framework within which this concept of STR is situated is found in the following elaboration of the general regularities of production and technical revolutions by two Soviet theorists, Shukhardin and Kuzin, which, for fear of doing it injustice, I have cited in full:

> Each socioeconomic formation has a characteristic system of technique and technological mode of production, and in each formation the producer occupies a definite place in production. The birth of elements of new technique takes place within the old mode of production. The establishment of the mode of production of the new formation passes through two stages. In the period of the first phase it uses the old material and technical basis inherited from the previous formation, and the old system of technique and techno-

logical mode of production still exist. (The production relations change, as does the mode of uniting the producers with the means of production.) As a result of the technical revolution and then the production revolution, the new material and technical basis is created and the new mode of production is victorious. The second phase means the full establishment of the mode of production of the new formation. (cited in Cooper, 1977, p. 155–156)

An important concept in the STR theory is the distinction between the technical and production revolutions. By production revolution is meant a broad process of "changes in the class structure of society, the formation of the industrial proletariat, and the establishment of the full domination of capitalist production relations" (Cooper, p. 154). The technical invention, on the other hand, refers to the preparation period between initial technical inventions and the production revolution during which the emergence of inventions leads to a revolution "in the means of labor, the forms of energy, production technology, and the general material conditions of the production process" (p. 154–155). The significance of such a distinction, according to Cooper, lies in that only production revolution can lead to fundamental changes in production relations, including the liberation of workers from their subordination to technology. According to proponents of the STR theory, the capitalist system only reinforces the social subordination of workers whereas the socialist system is the only system that can provide the context for production revolution, where "the new nonexploitative production relations [are] formed after the socialist revolution" (p. 158).

How, then, one might ask, is the socialist system superior to the capitalist regime in this regard? The answer lies in another important distinction the STR theory makes—that between subjective and objective elements of technology. The STR theory claims that technology developed in the last two centuries of the Industrial Revolution contains both subjective and objective elements. According to Dzherman Gvishiani (1972), a prominent Soviet theorist and also the deputy chairman of the USSR State Committee for Science and Technology, subjective elements refer to ownership functions associated with the use of technology that "implements and protects the interests of the capitalists," and objective elements refer to those management functions in the use of technology that are dictated by the nature of large-scale social production (p. 53). Under capitalism, claims Gvishiani (1972), the objective elements of technology cannot fully develop because it directly contradicts the subjective requirements of capitalist production. The quest for scientific and technological progress in the capitalist system is characterized by the sole motive of profit making. Therefore, "science and technology advance in an exceptionally contradictory way squandering their potential and ruling out any possibility of scientific co-operation on the scale of the entire society" (p. 40).

On the other hand, socialism, according to Gvishiani (1972), provides the most favorable objective conditions and an obstacle-free environment for science and technology development. Unlike capitalism, socialist management is not in contradiction with the organizational and technical aspects of technology use. On the contrary, it is in correspondence with the requirements of modern, large-scale production.

This prioritizing of socialist systems over capitalist systems, in the absence of a sound rationale, especially of the support of empirical evidence, is no doubt problematic. Underlying such an argument is the assumption that certain elements of technological development are universal, objective, and natural. Like the theory of technological determinism, the STR theory denies the historical and cultural nature of technology. An overly optimistic perspective, the STR theory assumes a natural compatibility between the capitalist technology and the Soviet goal to create a socialist society. As Cooper (1977) points out, although such a perspective does not deny the interrelationship between new technical means and organizational and other social changes, it does deny the fact that "the content and social meaning of these changes are uniquely determined by the fact that the innovations derive from a different social system" (p. 174). Such a perspective stems from two notions held by STR proponents. One is that technique, as "artificially created means of activity of people," is separate from the social consequences of its use; the other is that "technique as such . . . is regarded as neutral with respect to social systems and class" (p. 147). Understood from such a perspective, the distinction between "capitalist" and "socialist" technique is then no longer valid. This simplistic and overly optimistic view about "the positive role of science and technique in the development of society and for the liberation of mankind" (p. 175) is exactly what the problem is with the STR theory.

Compared with the first phase, one significant change about the technological theorization in this second phase is the acknowledgement of the impact of technology on society and culture. Even though the theories of technological determinism, substantive technology, and technological convergence are unduly pessimistic about man's role in technological change, and the theory of scientific and technological revolution is overly optimistic about socialist systems, these theories nevertheless represent progress toward a better cognitive understanding of technological phenomena and a better preparation for human interaction with technology.

Phase 3. Exploratory Interaction

As a natural development in the evolution of technological theorization, this third phase of technology theories, which includes the ambivalence theory and the mediation theory, features a move toward a more dialectic view of technological change. The theory of technological ambivalence was developed by Andrew Feenberg in his early stages of theory building about technology development and transfer. The basic tenet of the ambivalence theory is that "technique (and culture) is not neutral but ambivalent, capable of various alternative developments, of growing according to different class criteria" (Feenberg, 1975, p. 599). Accordingly, technological development and socialist cultural values are not necessarily incompatible, as Feenberg (1977) claims:

> Although created under the cultural and economic constraints of capitalism, the imported technology may be bent to new ones in a new context. It may then be routinely and efficiently employed in the service of cultural values quite different from those that presided over its creation. (p. 109)

Capitalist technology when transferred to a socialist society, according to Feenberg (1977), does not automatically become socialist technology. "The ambivalent nature of capitalist technology requires that class power be used to change the very nature of work and technology" (p. 475). Capitalist technology, therefore, goes through, in Feenberg's words, a "bootstrapping" process, which reshapes the inherited technology. It is then put to new uses in the socialist context and is employed to produce new technological means, which are fully adapted to the socialist culture.

Feenberg's (1977) argument of technological ambivalence is based on Marx's theory of transition to socialism, which he summarizes as follows:

1. Systems of control from above (class rule) require a division of labor and criteria of innovation incompatible with the full and democratic development of the individuality of the workers;
2. A system of workers' self-administration would require the development of a quantitatively different division of labor compatible with the employment of highly educated and socially responsible workers;
3. Such a self-administered system would be oriented toward long-range patterns of technological development that would further the ever-fuller actualization of human potentialities at work. (p. 91)

What happens in this process, according to Marx, is that the workers make use of the "ambivalent" means to meet their ends by (a) using inherited or transferred elements to consolidate their power, (b) transforming these elements over an extended period of time until they have built a radically different technological base that is adjusted to their own needs, and (c) using their class power to set standards and goals for their society so as to shape the particular ambivalent potentialities of the heritage that are conducive to their own culture (Feenberg, 1977, p. 109). Such a Marxist perspective regards transferred capitalist technologies not as given and completed entities but as starting points in a process of cultural transformation. Feenberg sees this developmental approach as the major difference that distinguishes his ambivalence theory from the neutrality theory.

Feenberg (1977) developed this theory as a response to the convergence theory with respect to the debate between convergence and transition on the issue of technology transfer. He challenges the deterministic notions of the convergence theory about the fixed patterns of and cultural subordination to technological development. He argues that on the issue of the "development paradox"—the contradiction between technology and human freedom—the question is not whether technology merely contributes to a process of socialist modernization convergent with that in the West, but "whether the transferred technology takes its place in a general process of cultural change leading to a different type of industrial society from that in which it originated" (p. 74).

This process of cultural transformation of capitalist technology, according to Feenberg (1977), is seen in the China case. In the face of the many problems associated with technology transfer and the gap it produces between individual cognitive capacities and the demands of technology, such as the transformation of economic life, instant knowledge deficit, and the rise of managerial control from above, China is "attempting to implement Marx's own remedy, the gradual abolition of the division of mental and manual labor, with, as a long range consequence, *the transformation of 'capitalist' technology into 'socialist' technology*" (p. 83, emphasis mine).

Feenberg's (1977) ambivalent theory is optimistic in nature, especially on the issue concerning the ability of socialist cultures to transform capitalist technologies. This is reflected in his definition of the process of transition:

> The transformation of capitalist technology is conceived as a lengthy process in which changing social relations would create the conditions for a democratic reproduction of the mechanical base. Presumably, new social criteria of innovation, responding to the interests of the producers, would prevail over the values embodied in capitalist technology. (p. 90)

Such an optimism also comes from the several "indices" that he sees in an emerging transitional process of technology transfer, which can be summarized as follows: (1) an alternative to capitalist reward structures that emphasizes "public models of consumption of goods and services and 'ideological' motivations for the acquisition and application of skills;" (2) social structure that insures workers' access to knowledge; (3) a reliance on "the exercise of authority by highly trained professional and managerial personnel," which contributes to "the enlargement of the workers' initiative and control throughout society;"and (4) means by which "workers can intervene in the activities of those with power and authority" (p. 97–98).

What, then, distinguishes ambivalence theory from the neutrality theory since both seem to allow different possibilities with the use of technology? First, neutrality theory takes transferred technology wholesale from one culture to another and assumes its applicability without question. On the other hand, ambivalence theory, as mentioned above, takes a developmental approach and situates the transferred technology in the context of the new culture and assigns to it a role in the new process of technology development, responding to a set of new conditions within the new cultural environment.

Second, the neutrality theory sees no inherent values in the technology itself, which, according to Fleron (1977a), is why the neutrality theory regards technology as neutral with respect to ends and social goals. Values are external to technology and are brought to bear on technology by users. The ambivalence theory, on the other hand, recognizes both the positive and negative elements of technology. Capitalist technology, according to the ambivalence theory, has both liberatory and repressive elements and exhibits mostly its repressive elements under the capitalist system. Therefore, the task of socialism is to repress those repressive elements and ensure that the liberatory elements predominate under the socialist system.

A third distinction, which is not so much a difference between the two as a unique feature of the ambivalence theory, is the political optimism of the nature of its claims about socialism's super ability to transform capitalist technology to serve its own ends: "that powerful liberating aspect of technology joins socialism's struggle to realize the new human freedom" (Fleron, 1977a, p. 471). Accompanying such an optimism about the socialist system is an implied, corresponding pessimism about the capitalist system. Either sentiment, however, is yet to be substantiated with empirical evidence.

The mediation theory is proposed by Fleron (1977) as an alternative approach to the neutrality theory, the theory of technological determinism, the theory of scientific and technological revolution, and the ambivalence theory. Fleron rejects these theories as either too pessimistic to allow humanity any hope of transcending technological problems (technological determinism), or too optimistic to account for the obvious problems encountered by socialist societies today related to the use of foreign technology (ambivalence theory and the STR theory), or too simplistic to explain the obvious connection between technology and culture.

A fundamental question, one that precedes the question of the impact of technology on society and culture but is ignored by these theories, according to Fleron (1977a), is the impact of society and culture on technology. The mediation theory takes as its starting point the need to understand "the extent to which a particular technology . . . in its formative stages was a material reflection, embodiment, or reification of dominant cultural values" (p. 472). Such an understanding, argues Fleron, will be able to provide us with clues about the impact of technology on other societies and cultures and the likelihood of those societies to develop transformed technologies and modified cultures of technology.

The basic argument of the mediation theory is that "technology as one of the artifacts of culture embodies the dominant values contained in that culture" (Fleron, 1977a, p. 472). As a product of a particular historical context, argues Fleron, the technology developed under capitalism is a material concretization, manifestation, and reification of capitalist ideologies and the goal of maximum control over the labor force in order to maximize profits. This control function is reflected not only in the physical technical object itself, but also in its accompanying forms of technical rationality and infrastructure, not only in its formative stages, but also in its later stages of refinement. This is not to say that the retaining of the control function and other elements of capitalist production relations and values in capitalist technology requires the permanent presence of the substructural and cultural reinforcement. Because of the deep extent of interpenetration between capitalist productive forces and cultural forms (such as technology), the capitalist superstructure, including cultural forms, which in turn includes technology, has become increasingly "dense," which means "it is very difficult to overcome the subjective (exploitative) elements of culture" (Fleron, 1977a, p. 473). Therefore, Fleron claims, when transferred from one culture to another, technology as a cultural form brings with it a particular cultural content that is hard to diffuse, and it will be equally hard to simply infuse a new cultural content into the old form (p. 474).

The purpose of the mediation theory, according to Fleron (1977a), is to answer such questions as:

> When will the repressive aspects of this capitalist technology be transcended?
>
> At what point in the development of socialism—in the period of transition—will the limitations of capitalist technology be overcome and human liberation thereby be made noticeable?

> What concrete indicators shall we look for, both in the theory and the practice of socialism, as signs of this transformation? (p. 475)

The ambivalence theory and the STR theory have not provided us with answers to these questions. The mediation theory, Fleron admits, does not have ready answers, either. However, it at least points to where the problem is. The following statement by John Hardt (1975) about the case of computer technology, suggests Fleron (1977a), may provide us with more clues in this regard:

> A reform in the system of planning and management is probably more important to performance than computer hardware is. [. . .]
>
> A sage Western observer once noted that the best approach to efficient computer usage would be to design the operation as if it were to be converted to computer application and then stop short of conversion. This seems to be the opposite of current Soviet thinking, which is likely to overemphasize the change in institutions required to make effective use of the computer technology. (cited in Fleron, 1977a, p. 479)

Fleron acknowledges that, as this statement reveals, technology involves more than hardware; "it includes broader cultural factors of infrastructure and rationality," and for the computer to function as it was designed, it requires the presence of these elements (p. 480). Fleron's own statement concerning cybernetics and automation is perhaps more revealing:

> If mechanization forced us to coordinate our bodies to the requirements of the machine, then surely cybernetics and automation force us to coordinate our minds to the requirements of the machine. The mediation theory of technology suggests that this will be the result of applying the cybernetics and automation developed under capitalism, since they, like every other level of capitalist technology, are vehicles for the mediation of capitalist cultural values and patterns. So tainted, this higher level of technological development can be

> no more liberating in a socialist context than were the earlier forms of capitalist technology. (p. 483)

Obviously, what Fleron (1977a) is pointing to is that the transfer of modern technology, as in the case of cybernetics and automation, requires our adjustment in both physical and mental capacities. The mental adjustment is required due to the capitalist cultural values and patterns that come with technology. Users in the target society obviously need to mediate such values and patterns to render the technology useful.

Phase 4. Proactive Control

If phase three above represents man's initial efforts at turning the tables, so to speak, in his reciprocal relationship with technology, phase four of technological theorization, which features such theories as the critical theory of technology, the rhetorical theories, and the social theories, signifies a more substantive attempt at gaining control over particular aspects of technological developments, though such control is never total and complete.

Possibly one of the most influential theories in the field of technology study, Feenberg's (1991) critical theory picks up exactly where Fleron's mediation theory left off: it attempts to provide an answer to the question of transformation of capitalist technologies in socialist contexts. Compared with the ambivalence theory that he developed in the 1970s, the critical theory represents a later stage in Feenberg's theory development about technology transfer. His *Critical Theory of Technology* (1991), published almost fifteen years after his articulation of ambivalence theory, presents a more elaborate, more critical perspective on technology. The basic argument of the critical theory is that technology is not a simple object, but "an 'ambivalent' process of development suspended between different possibilities" (Feenberg, 1991, p. 14). As a response to the technological deterministic propositions of fixed patterns of technological development and social and cultural submission to technology, the critical theory advances the following two counter propositions:

1. Technological development is over determined by both technical and social criteria of progress, and can therefore branch in

any of several different directions depending on the prevailing hegemony.
2. While social institutions adapt to technological development, the process of adaptation is reciprocal, and technology changes in response to the conditions in which it finds itself as much as it influences them. (Feenberg, 1991, p. 130)

The basis of these propositions is the notion that "technical objects are also social objects" (p. 130). The actualization of technology from among a variety of possibilities is achieved only when technology is placed into an actual context where technical and social determinations intersect. Only then can suspended potentialities of technology be realized and become concrete technology that can serve the needs of the new cultural context.

Therefore, the mission of the critical theory, claims Feenberg (1991), is to "invent a politics of technological transformation," to explain how transferred technology can be redesigned to adapt to the needs of the new society (p. 13). In other words, it is to provide a theoretical framework that can rationalize a formal means to place transferred technology at the intersection point of technical and social criteria. The key to accomplishing such a task is what Feenberg calls "instrumentalization."

According to Feenberg, there are two kinds of instrumentalization involved in technology development and transfer: primary instrumentalization and secondary instrumentalization. The problem with such theories as neutrality theory and technological determinism, claims Feenberg, is their overemphasis on the former to the exclusion of the latter. Primary instrumentalization is what characterizes technical relations in every society, though with variation in terms of emphasis, range of application, and social significance from one society to another. Capitalism tends to define technique as a whole in terms of the primary moments of decontextualization, calculation, and control—the primary qualities of technique. Technique thus defined encompasses only the primary instrumentalization and excludes other aspects of technique or treats them as nontechnical (Feenberg, 1991, p. 182).

What is lost, or suppressed, in such a treatment is the "integrative potentialities of technique"—the secondary qualities of technique—"that often compensate for some of the effects of the primary instrumentalization" (Feenberg, 1991, p. 182). By "integrative potentialities,"

Feenberg means that as technologies develop, they "often reappropriate some of the dimensions of contextual relatedness and self-development from which abstraction was originally made in establishing the technical object relation" (p. 182).

This concept of integration leads Feenberg (1991) to argue that technique is dialectical and to advance the notion of a "secondary instrumentalization" (p. 182). A full definition of technique, argues Feenberg, must include both primary and secondary instrumentalizations. This secondary instrumentalization, explains Feenberg, "involves a reflexive metatechnical practice that treats finished technical objects and the technical relationship itself as raw material for more complex forms of technical intervention" (p. 182–183). In other words, it addresses the secondary attributes of technology and uses them as correctives to the primary attributes. Because technology transfer is a cultural and social event, it inevitably involves nontechnical as well as technical aspects, and secondary instrumentalizations are what integrate the technical and non-technical aspects of technology. The transition from primary to secondary instrumentalizations, Feenberg points out, is a normal phenomenon in the process of technology development and transfer.

To understand this concept of instrumentalization further, let's take a more operationalized look at Feenberg's definitions of primary and secondary instrumentalizations. According to Feenberg (1991), primary instrumentalization includes the following four aspects:

1. *decontextualization:* the separation of object from context;
2. *reductionism:* the separation of primary from secondary qualities;
3. *autonomization:* the separation of subject from object; and
4. *positioning:* the subject situating itself strategically. (p. 184)

Decontextualization is the act of artificially separating technical objects that are reconstructed from natural objects from the systems and contexts where they originated. Once decontextualized, these objects are then fragmented into various parts and analyzed as such. Such an isolation detaches the technical object from the particular role it plays in nature. "Technology is constructed from the bits and fragments of nature that, after being abstracted from all specific contexts, appear in a technically useful form" (Feenberg, 1991, p. 185). Under capitalism, according to Feenberg, the technical skills of human workers are decontextualized in this manner. In order to get at the technical elements of the workers, workers are isolated from institutions such as family and community and reduced to instrumentalities. Workers on the assembly line, for example, are treated neither as members of a community that play natural roles, nor as slaves that provide pure manual labor power; instead, they are treated as components of machinery (p. 185).

Reductionism is the effect of decontextualization, by which primary qualities are separated from secondary qualities. The capitalist control of the labor force, or management as we commonly call it, relies on technical means supplied by the decontextualization of objects. These technical means are abstracted from concrete natural social objects by means of "a reduction of complex totalities" to those elements that are subject to capitalist control (Feenberg, 1991, p. 185). Lost in this reduction process are those untransformable elements that associate the object with "its pretechnical history and its potential for self-development" (p. 185–186). Those controllable elements are what Feenberg means by "primary qualities," and those untransformable elements are the secondary qualities. A full examination of any technical object must include both its primary and secondary qualities. A reduction of it to the former is likely to result in neutrality or deterministic perspectives.

Autonomization refers to the means by which the subject is isolated from the object by dissipating or deferring the impact of the object on the subject. This essentially denies the cause-and-effect relationship and the action-reaction law that inevitably exists between the subject and the object in any given action. It assumes a one-way action course by which the subject acts on the object without being affected at all by the reaction of the object. In capitalist production relations, this has significant implications as the subject (manager) and the object (worker) are both human beings. In the case of technology use, one cannot approach any technical system without being affected one way or another by it (Feenberg, 1991, p. 186–187).

Positioning is the strategic situating of the technical subject in the context of and based on the basic law of its technical objects. The purpose of the subject is not to replace the law but to modify it. Under capitalism, the capitalist strategically establishes a position where he acts on the social reality rather than out of it. He positions himself external to the system in which the technical objects functions. "Capitalist practice thus has a strategic aspect: it is based not on a substantive role *within* a given social group but rather on an external relationship to groups in general" (Feenberg, 1991, p. 188, emphasis original).

When applied to the organization of labor, argues Feenberg, these four primary qualities produce an alienated system. Technology is reduced to some universal qualities at the cost of its social and cultural attributes and is inevitably treated as a decontextualized, acultural object that provides tools of control for the capitalist. This is why, according to Feenberg (1991), we need a secondary instrumentalization, which includes four moments:

1. *concretization,* the recontextualizing of the decontextualized technology;
2. *vocation,* the recovery of the reciprocal relationship between subject and object;
3. *aesthetic investment,* the recovery of the secondary qualities of technology; and
4. *collegiality,* a new, more cooperative relationship between management and the labor force. (p. 189)

Concretization is "the discovery of synergism between technologies and various environments" (Feenberg, 1991, p. 189). Because capitalist decontextualization deprives technology of its contextual elements in its class-specific application to serve capitalist interests, a more democratic use of technology requires, in Feenberg's terms, a "recontextualizing practice" that will represent the wide range of interests that capitalism represents only partially, and to recover those interests that reflect human and natural potentialities ignored and suppressed by capitalism (p. 189). This concretization practice is not incongruent with the nature of technology because, as Feenberg argues, "*the passage from abstract technical beginnings to concrete outcomes is a general integrative tendency of technological development* that overcomes the indifference of the reified elements and spheres characterizing the heritage of capitalist industrialism" (p. 194, emphasis original).

Vocation, the acquisition of craft, is used here by Feenberg to mean the de-isolation of the subject from the object and the recovery of the reciprocity of the relation of subject to object. Here, the subject no longer stands aloof above the object, but instead "is transformed by its own technical relation to them" (Feenberg, 1991, p. 190). This essentially recognizes the reciprocal relationship between culture and society on the one hand and technology and its use on the other.

Aesthetic investment refers to the recovery of the secondary qualities of technology lost in primary instrumentalization and the reinsertion of "the object extracted from nature into its new social context" (Feenberg, 1991, p. 190). Aesthetic investment is what counters reductionism in the primary instrumentalization mentioned earlier.

Collegiality refers to "the praxis of voluntary cooperation in the coordination of effort" (Feenberg, 1991, p. 190). It is Feenberg's proposed substitute for control advocated by capitalism. A collegial relationship between management and the labor force can only result in a fuller representation of the interests of all parties involved in technology development. It has, according to Feenberg, the potential of reducing alienation by replacing control from above with conscious cooperation.

The significance of these four qualities lies in their ability to mend the harm caused by primary instrumentalization of technology. They make up an essential stage in technology development which is often overlooked by many theorists of technology, especially those who hold neutral and deterministic views. As Feenberg (1991) puts it,

> These four dimensions of technique are higher-order properties associated with the dynamics of socio-technical systems. They support the reintegration of object with context, primary with secondary qualities, subject with object, and leadership with group. (p. 190)

Such a secondary instrumentalization, according to Sullivan & Porter (1997), is "helpful to our thinking" and "can be usefully applied to critical research" (p. 105). In operational terms, *concretization* means to "contextualize and situate both technologies and participants"; *vocation* to "treat participants and technologies holistically, not reducing either simply to constituent elements (variables), but respecting the complexity of their subjectivities"; *aesthetic investment* to "exhibit a concern for overall quality of life/work for participants"; and *collegiality* to "work with participants in 'the coordination of effort' and share responsibility for improving institutional and social conditions and work conditions" (Sullivan & Porter, p. 106). Understood this way, Feenberg's secondary instrumentalization enables the transition from reification, the capitalist abstraction and decontextualization, to reintegration, the socialist recontextualization of technology. It provides a critical missing link in the development process of technology.

Perhaps, critical theory is best understood in terms of its similarities to and differences from neutrality theory and the theory of technological determinism. Although it does agree with the theory of technological determinism in arguing that technology goes beyond being a sum of tools and does shape society and culture in an "autonomous" fashion, it rejects the notion that technology dictates in all societies the type of "atomistic, authoritarian, consumer-oriented culture" typical of Western, capitalist societies (Feenberg, 1991, p. 14). The contextualized nature of technology prohibits the prescription of one single "technical phenomenon" that can characterize the development paths of all technologies in all cultures, especially in socialist societies.

Like neutrality theory, critical theory rejects the fatalism of technology perspectives of Ellul and Heidegger. It is neither overly pessimistic nor overly optimistic in the face of technology development but acknowledges different possibilities with technology. However, the differences between critical theory and neutrality theory far exceed their similarities. Critical theory rejects the notion of technological neutrality and argues instead that technology is embedded with the values of the culture where it originates. Cultural values and class interests permeate the very design of technological procedures and rationality. The dominant form of technological rationality "stands at the intersection between ideology and technique where the two come together to control human beings and resources in conformity with what I will call 'technical codes'"(Feenberg, 1991, p. 14). Unlike neutrality theory, which sees only the primary instrumentalization of technology, critical theory sees secondary instrumentalization of technology as an essential stage that completes the process of technological development. As pointed out earlier, the critical theory of technology recognizes technology's integrative potential and exposes the obstacles to its release; it "thus serves as the link between political and technical discourse" (Feenberg, 1991, p. 183).

Despite all the differences between the theories of technology outlined above—some are technically oriented, such as the theory of technological determinism; some are culturally oriented, such as the STR theory, the ambivalence theory, and the critical theory—one thing seems to be common to them all: they seldom regard technology development as a rhetorical act of the construction and reconstruction of meaning about technology, and even when they do, their perspective is often a pessimistic one. Richard Ohmann (1985), for example, while acknowledging the rhetorical nature of technology development, views technology as a capitalist instrument for perpetuating its hegemony. "Computers," he argues, "are a commodity" designed "to facilitate the marketing of still more commodities" and therefore only lead to "new channels of power through which the few try to control both the labor and the leisure of the many" (p. 684). Even though he grudgingly acknowledges the possibility of monopoly capital to "generate resistance and rebellion," his conclusion about new technologies such as the computer is a gravely pessimistic one:

> But this age of technology, this age of computers, will change very little in the social relations—the class relations—of which literacy is an inextricable part. Monopoly capital will continue to saturate most classrooms, textbooks, student essays, and texts of all sorts. It will continue to require a high degree of literacy among elites, especially the professional-managerial class. It will continue to require a meager literacy of none from subordinate classes. And yet its spokesmen . . . will continue to kvetch at teachers and students, and to demand that all kids act out the morality play of literacy instruction, from which the moral drawn by most will be that in this meritocracy they do not merit much. (p. 687)

Unlike Ohmann, however, most proponents of the social and rhetorical perspectives on technology take a more dialectic approach and attribute a more significant role to the context and the participants in technology development. This is where the true social rhetorical theories distinguish themselves from the above theories. They address technology development and use from social and rhetorical perspectives by situating the participant in the center of the act and view technology development as a process of social construction and discursive formations. One thing is important to note here: though a social approach is not necessarily rhetorical, rhetorical approaches are inevitably social. This is why I have chosen to discuss these two seemingly distinct approaches together.

In investigating the situated practice concerning technology, Lucy Suchman (1987) examined the situated actions and practices of users of a copying machine. She sees machine actions as "determined by stipulated conditions" and the machine interaction with the world and the people as "limited to the intentions of designers and their ability to anticipate and constrain the user's actions" (p. 189). Such a view recognizes the technical rationality of technology and sees technology as enculturated. People's use of technology, therefore, is context driven. Suchman concludes that "insofar as actions are always situated in particular social and physical circumstances, the situation is crucial to action's interpretation" (p. 178). What knowledge means and how it is constructed, she contends, must be understood in the context of the particular circumstances in which knowledge constructions occurs. Thus, "the contingence of action on a complex world of objects, artifacts, and other actors, located in space and time, is no longer treated as an extraneous problem with which the individual actor must contend, but rather is seen as the essential resource that makes knowledge possible and gives action its sense" (p. 179). Such a perspective is rhetorical in nature though Suchman does not use the explicit term.

Like Suchman, Terry Winograd and Fernando Flores (1986) take a social and situated approach to computer design. They argue for an ontological design, which grows "out of our already-existent ways of being in the world, and deeply affecting the kinds of beings that we are" (p. 163). They see a reciprocal relationship between technology and people, in that technological development leads to "a changing awareness of human nature and human action, which in turn leads to new technological development" (p. 163). On the one hand, computer systems, for example, reinforce our rationalistic interpretation of human action, and "working with them can reinforce patterns of acting that are consistent with it" (p. 178). On the other hand, "we can create computer systems whose use leads to better domains of interpretation. The machine can convey a kind of 'coaching' in which new possibilities for interpretation and action emerge" (p. 179). Technology, understood in this fashion, is both a reinforcement of tradition and "a vehicle for the transformation of tradition" (p. 179).

Harrod Innis, in his 1951 study, *The Bias of Communication,* also situates technological development in social and historical contexts. He argues that "if we are to understand as fully as possible the transformations brought about by complex technological developments, especially in the realm of communications, then we must have a sense of the way things were before such transformations" (cited in Heyer & Crowley, 1991, p. xvii). Culturally situated, knowledge construction (and its dissemination), about technology and other social phenomena, suggests Innis (1951), is basically a communication act: "A medium of communication has an important influence on the dissemination of knowledge over space and over time and it becomes necessary to study its characteristics in order to appraise its influence in its cultural setting" (p. 33).

This social nature of the medium of communication is also explored in Marshall and Eric McLuhan's 1988 study, *Laws of Media: The New Science*. Their investigation rests on two fundamental assumptions: (1) "each of man's artifacts is in fact a kind of word, a metaphor that translates experience from one form into another," (2) all [things] are equally artifacts, all equally human, all equally susceptible to analysis, all equally verbal in structure" (p. 3). By this account, technology is an artifact and therefore nothing but a translation of human experience. Such a perspective is not only social but very much rhetorical.

Langdon Winner's approach to technology, although not exactly rhetorical, is also social in nature. As we engage ourselves in technological development, Winner (1986) argues, we should ask this question: "As we 'make things work,' what kind of *world* are we making?" (p. 17, emphasis original). Winner suggests that "we pay attention not only to the making of physical instruments and processes . . . but also to the production of psychological, social, and political conditions as a part of any significant technical change" (p. 17). In effecting technological change, we should "design and build circumstances that enlarge possibilities for growth in human freedom, sociability, intelligence, creativity, and self-government" (p. 17).

Such social approaches point to factors within social and historical contexts that affect and are affected by technological change. In recognizing such a reciprocal relationship between technology and society, rhetorical approaches go one step further to examine how technological developments are acts of discursive formations. A good starting point in understanding the rhetorical perspective is perhaps Walter Ong's argument that writing itself is technology. Writing, as a means of separating the speaker from the spoken word, argues Ong (1977), calls for "massive technological interventions" and requires "reflectively prepared materials and tools" (p. 22). Contrary to oral speech, writing is an artificial act, just like any other acts of using technology. Technologies, Ong (1982) further contends, "are not mere exterior aids but also interior transformations of consciousness, and never more than when they affect the word" (p. 82), and when "properly interiorized, [technology] does not degrade human life but, on the contrary, enhances it" (p. 83). The interiorization of technology and the representation of this interiorization through discourse are what make technological development and use a rhetorical phenomenon.

Take the computer for example. When used properly, computers can provide a democratic platform of exchange in the classroom, according to Thomas Barker and Fred Kemp (1990). They contend that networked computers can be used to promote "a new model of classroom interaction" that empowers "the student writer and deneutralizes his sense of text" (p. 26). With the use of social-constructionist models, computers can also serve as the locus for collaboration that will "encourage the open expression of diversity and the active participation of all students" (p. 26).

Cynthia Selfe and Richard Selfe's (1994) examination of computer interfaces is even more rhetorical in nature. Computer interfaces, argue Selfe and Selfe, are rhetorically constructed and function rhetorically in several ways. As maps of capitalism and class privilege, "computer interfaces present reality as framed in the perspective of modern capitalism, thus, orienting technology along an existing axis of class privilege" (p. 486). As maps of discursive privilege, they increasingly systematize "the orientation of the interface along the axis of class privilege . . . by the application of related discursive constraints," for example, by privileging English as "the language of choice or default" (p. 488). As maps of rationalism and logocentric privilege, computer interfaces perpetuate an orientation "aligned with the values of rationality, hierarchy, and logocentrism characteristic of Western patriarchal cultures" (p. 491). Despite such rhetorical and cultural constraints imposed upon us by computer interfaces, however, Selfe and Selfe's perspective on computers is not deterministic. They advocate active participation by users to counter the rhetorical constraints through rhetorical means: to become critical of technology, to contribute to technology design, and to reconceive the map of the interface.

Sullivan and Porter's (1997) study takes a not only rhetorical but more situated and contextualized perspective on computer technology. They "argue against viewing computer technology and research participants in a detached, decontextualized way and in favor of seeing technology from a critical viewpoint as created by, situated in, and constitutive of basic human relations" (p. 102). Such a perspective "sees technology not as abstracted or decontextualized systems . . . but rather as involving real people using human-designed machines for situated purposes" (p. 102). Such a perspective places participants in the center of computer, and in fact any technology, development. A rhetorical perspective is not truly rhetorical if it does not recognize the central role of the participants in developing a technology.

Robert Johnson's (1998) study is a good illustration of such a perspective. Johnson takes an interdisciplinary approach and presents a model of what he calls the "user-centered rhetorical complex of technology." Working from a historical point of view and applying rhetoric theory, Johnson elaborates on this user-centered rhetorical theory of technology by defining three focal issues: user knowledge, human-technology interaction, and technological determinism. This "complex," according to Johnson, situates the "end" of technology in users (p. xv), which is what makes his perspective a rhetorical one.

Like the proponents of some of the other theories discussed above, Johnson acknowledges the dilemma that has "trapped" us in the face of technology development: we are both enamored and threatened by the seemingly endless power and possibilities of technology. In our struggle to find a way to take advantage of the great potentialities of technology while exerting control over the direction and speed of technological development, we are constantly pressed by the question: how can we, as users, have an effect on the course of technological development?

The answer, argues Johnson (1998), lies in "a refiguring of the *end* of technology: a fundamental rethinking of where technology is going and how humans can monitor its speed and direction" (p. 20). In modern history, the *end* of technology has been traditionally placed in the following:

1. The interest of the developers who hope to gain from it;
2. The interest of the disseminators who likewise hope to reap the fruits of its success; or
3. Those who develop and then release a technology into the public sphere with little or no concern about its *end* whatsoever. (p. 20–21)

What is left out of this list is one of the most important groups of individuals involved in technology development: the users, which prompts Johnson to argue that the end of technology should be refigured in the user: people (basically all us) who use various technologies on a daily basis (p. 21).

This exclusion of users, according to Johnson (1998), is largely attributed to the traditional system-centered approach to technology that has characterized technology design over the last two centuries. This approach "holds that the technology, the humans, and the context within which they reside are perceived as constituting one system that operates in a rational manner toward the achievement of predetermined goals" (p. 25). In this approach, the user is "far removed from the central concern of the system or interface design" (p. 28).

As an alternative to the system-centered approach, Johnson (1998) proposes a user-centered view of technology. The user-centered theory regards the user "as an integral, participatory force in the process," places the user at the center of the model, and replaces the system and the designer's image of the system with the user as the dominant feature of technology (p. 30–31). This approach also has a dimension that the system-centered approach lacks—the user's situation. The user's situation is defined as the activity the user is engaged in, such as learning, doing, and producing, and takes into account such factors as the tasks and actions the user will perform as a result of a particular context.

Apart from the user's situation, the basic components of the user-centered model are the same as those in the system-centered model, namely, the designer's image, the user, the artifact/system, and the interface. What is different, however, is how these components interact. In the user-centered approach, the users are active participants at all stages of technology development, including design, development, implementation, and maintenance. This means that "they are allowed to take part in a *negotiated process of technology design, development, and use* that have only rarely been practiced" (Johnson, 1998, p. 32). Within such a model, the designer receives input from all parties involved in the process—the user, the interface, the artifact, and the user's situation—and truly interacts with the user. The technological design so created is the result of a true negotiation. "Thus, collaboratively, the technology is created through a process of 'give and take' that places users on a par with the developers and the system itself: a space within which users and developers can learn to value each other's knowledge and accept the responsibilities of technological design and development in new, shared ways" (Johnson, 1998, p. 33). Such a view of technology development as a process of negotiation and construction of meaning is what makes the user-centered model a rhetorical perspective.

Another significant rhetorical perspective of technology, one that has provided much of the theoretical basis for my own rhetorical perspective, is the one advanced by Stephen Doheny-Farina in several of his studies of the last few years, most notably his 1992 book, *Rhetoric, Innovation, Technology: Case Studies of Technical Communication in Technology Transfers*. An elaborate discussion of his rhetorical perspective is done in Chapter 3. However, I will provide the gist of this theory here.

Doheny-Farina (1992) defines technology transfer as a "series of personal constructions and reconstructions of knowledge, expertise, and technologies by the participants attempting to adapt technological innovations for social uses" (p. ix). It is a reciprocal process of interpretation, negotiation, and adjustment, whereby the designers, the technology, and the users are all changed (p. 6). Like Feenberg and Johnson, Doheny-Farina holds that technology and its development process reflect social and cultural practices and their embedded values. The transferred technology is more than a product; it embodies a set of practices that are unique to the original culture and often incompatible with the practices of the target culture.

Unlike Feenberg, Doheny-Farina (1992) goes one step further to argue that technology transfer is a rhetorical process of "social, organizational, institutional interactions, interpretations, and negotiations" whereby every participant along the way constructs and reconstructs the technology based on his own experience (p. 7). Understood this way, technology transfer becomes essentially a rhetorical phenomenon. The barriers to successful technology transfer, then, must also be rhetorical, and indeed they are; according to Doheny-Farina, what constitutes the central rhetorical barrier to technology transfer is the differences in perceptions by the participants about the meanings of technology, about what constitutes the knowledge about technology. Therefore, the issues to address in technology transfer would be such rhetorical issues as the difficulty in discovering and communicating key knowledge from the designers of technology, the difficulty in establishing ideal examples based on users' experiences, the difficulty in establishing an effective level of detail that fits the needs of various participants, and the disparity of technologies and users across the transfer.

A point of significant progress in this phase four of technology theory building is that we have moved from passive reaction to and interaction with technology on to proactive control over particular aspects of technology development, however partial and indirect this control might be. A second significant development is the recognition of the shaping impact of the cultural scene on technological change. Such a recognition is nothing short of being revolutionary in that it has enabled a fundamental shift in our perspectives of technology from debilitating pessimism to cautious optimism, which makes human intervention possible.

However, one rather glaring gap in this theory repertoire is the lack of an operationalized model that will lead to meaningful and effective analysis of any technological development, without which this phase four of theory development is incomplete. It is precisely the purpose of my next chapter to identify an applicable framework to operationalize this body of theories by modifying and expanding Kenneth Burke's model of rhetorical analysis—the pentad.

3 Rhetoricizing and Operationalizing Technology

Due to the rich variety of elements involved in the transfer and development process of technology and its complexities, there are probably as many ways of categorizing technology transfer and development as there are people attempting to categorize them. Since I have defined technology transfer as a process of rhetorical construction, I will draw on certain concepts from Kenneth Burke's rhetoric to build my rhetorical model of analysis.

I choose to use Burke's concepts for my model of analysis because of an apparent parity between my project and the Burkean rhetoric. While technology transfer is a complex process that involves economical, social, cultural, philosophical, and many other aspects, my approach to the study of this process is mainly rhetorical. In comparison, while Burke's system of concepts covers such disciplines as philosophy, literature, sociology, and linguistics, "[h]is primary focus, however, could be considered a rhetorical one" (Foss, Foss, & Trapp, 1985, p. 153). Such a focus, however, does not imply a juxtaposition of approaches, that of the rhetorical with the philosophical, the cultural, the economic, etc., to the effect of choosing one to exclude the others. Nor does it place the rhetorical in opposition to the other aspects. Rather, what I mean by a rhetorical approach is that the technology transfer and development process will be examined in terms of its philosophical, cultural, economic, social (and whatever) aspects from a rhetorical point of view.

BURKE'S PENTAD/HEXAD

A central tenet in the Burkean rhetoric is that language constructs reality (see, for example, Burke 1952, 1953, 1954, 1961, 1962, 1966, 1969, 1972, 1989). It follows then that language and thought can be treated primarily as modes of action. This assumption is the foundation upon which Burke builds his approach—dramatism—to study human action and motivation. But before we discuss this dramatism, let's first take a look at Burke's definition of human action.

The human being, according to Burke, is composed of two things: "animality—the biological aspect of the human being that corresponds to motion, and symbolicity—the neurological aspect of the human being that corresponds to action" (Foss, Foss, & Trapp, 1985, p. 166). Animality is characterized by motion, the natural bodily processes such as growth, metabolism, and digestion. Symbolicity, the ability to acquire language or a symbol system, is where action occurs. Action, as opposed to mere motion, is symbolic as it requires "a terrific lot of verbalization" (Burke, 1966, p. 28). Action such as theorizing, planning, etc. are driven by "complex, alembicated purposes"—"aims developed by custom, education, political systems, moral codes, religions, commerce, money, and so on" (p. 28).

Action, according to Burke, has three basic conditions: freedom to act, purpose/will, and motion (Foss, Foss, & Trapp, 1985, p. 166–167). While motion is a necessary condition for action, what sets action apart from motion is the first two conditions—freedom and purpose, which do not exist in mere motion. These two conditions constitute human motivation, which prompted him to study human action through a perspective called dramatism:

> The titular word for our own method is "dramatism," since it invites one to consider the matter of motives in a perspective that, being developed from the analysis of drama, treats language and thought primarily as modes of action. The method is synoptic, though not in the historical sense. (Burke, 1952, p. xvi)

So, why dramatism? It is because, according to Burke (1989), this dramatistic method can "show that the most direct route to the study of human relations and human motives is via a methodical inquiry into cycles or clusters of terms and their functions" (p. 135)? In Burke's (1966) view, human action is drama, in the literal rather than metaphorical sense:

> In this sense, man is defined literally as an animal characterized by his special aptitude for "symbolic action," which is itself a literal term. And from there

> on, drama is employed, not as a metaphor but as a fixed form that helps us discover what the implications of the terms "act" and "person" *really are*. Once we choose a generalized term for what people do, it is certainly as literal to say that "people act" as it is to say that they "but move like mere things." (p. 55)

While "Burke's dramatistic perspective studies language as symbolic action rooted in motion and characterized by freedom and purpose" (Foss, Foss, & Trapp, 1985, p. 168), the method he has developed to study the motivation in symbolic action through this perspective is his pentad, a set of five terms—*act, agent, agency, scene,* and *purpose*—as a "grammar" for examining human motivation.

Burke says that whenever we describe a situation(i.e., what people are doing and why they are doing it), "any complete statement about motives will offer *some kind* of answers to these five questions: what was done (act), when or where it was done (scene), who did it (agent), how he did it (agency), and why (purpose)" (1952, p. x). Taken at face value, the Pentad might look a little like the basic principle of 4Ws and an H (*who, what, when, where,* and *how*) in journalism writing. However, the significant difference lies in what I would call "selective (de)emphasis" of the five terms in the use of the Pentad. By "selective (de)emphasis" is meant that different people will emphasize or deemphasize certain aspects of the Pentad depending on their situation and purposes. As Joseph Gusfield (1989) points out,

> But for Burke, a culture, like a playwright, seizes on some parts of the Pentad and deemphasizes others. Sociologists, for example, are far more likely than psychologists to emphasize scene rather than agent. They utilize a paradigm that searches for explanations of action in institutional features, in the defined situation rather than the personality of the actor. So, too, a more "sociological" playwright like Arthur Miller will draw his audience toward understanding his characters as responsive to institutions, to the context, the scene. (p. 15)

However, emphasis or de-emphasis in the use of the Pentad does not result in the dismissal of any of the five elements as all these elements are entailed in any given rhetorical situation and they are intrinsically interrelated. The existence of one of the elements necessarily entails that of the others. Burke (1989) explains this interrelationship as follows:

> [. . .] for there to be an *act*, there must be an *agent*. Similarly, there must be a *scene* in which the agent acts. To act in a scene, the agent must employ some means, or *agency*. And it can be called an act in the full sense of the term only if it involves a *purpose*. (p. 135)

To use Burke's terms, all these elements are cosubstantial with one another. The term he uses to describe this cosubstantial relationship is "ratio." There are ten ratios: scene-act, scene-agent, scene-agency, scene-purpose, act-purpose, act-agent, act-agency, agent-purpose, agent-agency, and agency-purpose. Reversal of the elements in each pair gives us another ten ratios. These ratios "allow for a more detailed examination of the various relationships among the terms" (Foss, Foss, & Trapp, 1985, p. 170).

> The significance of these ratios lies in that in a particular ratio, the first element by its very nature dictates to some degree the qualities of the second element. In the scene-act ratio for example, "the scene contains the act" in that "the nature of acts . . . should be consistent with the nature of the scene" (Burke, 1952, p. 3). In other words, "the scene is a fit 'container' for the act, expressing in fixed properties the same quality that the action expresses in terms of development" (p. 3). "A church scene, for example, determines that only certain acts with certain characteristics will be performed there. Praying . . . would be proper in the scene, while doing cartwheels would not" (Foss, Foss, & Trapp, 1985, p. 171).

On the other hand, in a reversal of this pair of elements, whereby we obtain an act-scene ratio, the nature of the act determines that it could only happen in scenes of particular characteristics. An act of political speech, for example, could appropriately take place in a big conference room scene, but not in a department store. In an agent-act ratio, the agent's character requires acts consistent with his/her character. For a hero, for example, the act of saving somebody's life would fit the expected qualities of his character whereas murdering an innocent person would not.

These five elements and the different ratios, according to Burke, are ubiquitous, "for they are at the very centre of motivational assumptions" (1952, p. 11). In most cases, they do not appear in their literal terms but assume instead many guises "in the various casuistries" (p. 11). In general contexts, for example, terms such as "society" and "environment" are synonyms for "scene." In philosophical contexts and everyday speech, the word "ground," which is often used to discuss motives, also denotes the scene. In political contexts, "situation" is often used to indicate the scene, and a particular policy as a result of a particular social context is an example of the scene-act ratio in that the social context determines that such a policy would be made (p. 11–12).

One implication of these ratios that is of particular relevance to my study is the concept of "selective (de)emphasis" that I mentioned earlier. As we construct realities through linguistic means, we examine rhetorical situations with culturally tainted perspectives, as a result of which we strategically select, consciously or unconsciously, the elements for emphasis or deemphasis. The resulting reality we construct is thus only a selective interpretation of it. As Burke (1954) points out,

> We discern situation patterns by means of the particular vocabulary of the cultural group into which we are born. Our minds, as linguistic products, are composed of concepts (verbally molded) which select certain relationships as meaningful. Other groups may select other relationships as meaningful. These relationships are not *realities,* they are *interpretations* of reality—hence different frameworks of interpretation will lead to different conclusions as to what reality is. (p. 35)

Such selective interpretation is determined by "the particular linguistic texture" we are born into, though to some extent we can manipulate that linguistic texture (p. 36).

As Haixia Wang (1993) points out, two things are worthy of note here.

> First, while the sophistic rhetoricians believe that "deception" in the sense of choosing what to believe as truths or what to accept as one's own opinion is indispensable, Burke more explicitly points out, further, that both the rhetor and the audience are always already culturally/ideologically trained, forever incapacitated as well enabled by their cultural concepts. In fact, it is these cultural concepts and ideologies that make the communication between the rhetor and the audience possible in different persuasive modes or style, and it is the same concepts and ideologies that facilitate the arousing and fulfillment of desires and interests. (H. Wang, 1993, p. 19)

Another point, argues Wang (1993), is that while "deception" is considered inevitable, Burke considers it also "fundamentally ethical" (p. 19). The act of choosing and deciding implies the use of the will. Even the act of choosing to be shaped by a particular linguistic texture is willed. As Wang points out:

> The willingness and the ability to choose and to accept one opinion over another is indeed based on the fundamental sophistic notion that through language human beings can only deal with probable truths. But his sophistic argument is furthered by Burke's argument of its being fundamentally ethical by viewing language both as the source of human conflict and as the hope of human beings' overcoming it. Choosing, deciding, and willing make possible imagination, creativity, discussion, and correction in the human drama. (p. 19–20)

According to Wang, this dramatistic view of language, which views language and thought primarily as modes of symbolic action, is important in that discursive practices are inevitably pervaded by ideology and need to be considered in their social and historical contexts and that "an understanding of what happened in and to a discourse requires as thorough as possible an understanding of the cultural concepts which enable as well as participate in the communication between the rhetor and the audience" (p. 20). This essentially points to the argument that culture is a player in discursive (and in fact all other) practices and that one's act of using a language entails using and being used by the value-laden linguistic texture at the same time. In the case of technology development, this means that while the individual perspective on technology contributes to the overall cultural perspective, it is also shaped by the cultural perspective. Thus, whatever meaning is constructed of a particular technology is a result of negotiation between the individual and collective voices within that culture.

I will now turn to the five elements of the Pentad and explain how they provide the methodological framework for my rhetorical model of analysis, in other words, what these elements translate into in the context of writing technology development.

Scene—Exigency

Scene, according to Burke (1952), can be viewed as the "container" of the act when used in the sense of setting, or background (p. 3). It is "the background of the act," "the situation in which it occurred," or "when and where it was done" (p. x). In the case of writing technology development, what scene translates into is the social context, which includes various aspects such as the economic, the historical, the political, the cultural, etc. Broadly speaking, it is the big environment/context in which the technology is developed. Now, how this environment relates to the act of technology development is a matter of the interrelationship between the scene and the act. Using Burke's terms, we are in fact talking about the act-scene ratio and the scene-act ratio. Let's look at these two ratios separately to see what clues they give us about the factors in the context/environment that influence technology development.

In the act-scene ratio, Burke tells us that the nature of the act determines that the scene must possess certain features for the act to take place. In other words, for the act of technology development to be possible, there must be certain elements in the social context that are conducive and essential to the development. One thing stands out as a prerequisite to any technology development—a need. However, that a need exits does not mean something will necessarily happen. Only when a need acquires a high degree of urgency and magnitude will it effect that development. Hence, the term "exigency" seems more appropriate as it better captures the essence of the kind of need that makes things happen.

Considered from a different perspective, that of the scene-act ratio, we see the same thing. According to Burke, the scene-act ratio indicates that the nature of a particular scene determines that acts of a particular nature will happen. In other words, when certain conditions in a particular scene are right, certain acts will be inevitable. This means, in the case of technology development, that when the social context acquires particular conditions, technology development will inevitably take place. Therefore, if an exigency—a pressing need for a particular technology—exists, that technology will be developed. This exigency is of various natures: it can be economic, political, or ideological, and more often than not it is a combination of two or more of these aspects, although one of them might be more dominant. For example, many countries adopted the automobile technology in the early part of the twentieth century mainly due to an economic exigency (to boost their economy by strengthening their transportation system, for example), although some countries also had secondary motives, such as to improve political status with an improved economy. On the other hand, most countries advance their military technologies to build a strong military force so as to achieve more leverage in political negotiations on the international scene. Whatever the reason, such an exigency has to exist for technology development to occur, although this exigency may be more obvious at some times while less apparent on other occasions.

Agent—Participants

Agent, in Burke's Pentad, is the person who performs the act. It is the people involved in the act. It follows then that the agents in technology development are the participants of the process. Two things are worthy of note here. First, not all people involved in an act are necessarily agents. People are of two different roles in an act. On the one hand, there are people who actively participate in the act, shaping the course of action. These people are the agents of the act. On the other hand, there are people who play marginal roles whose actions do not directly shape the course of action but instead form part of the context, the scene. Only people who play an active role in the act are agents. These people are the real participants of the act.

Second, the term *agent* is a rather inclusive term. It denotes a range of different people. As Burke (1952) points out:

> Meanwhile, we should be reminded that the term *agent* embraces not only all words general or specific for person, actor, character, individual, hero, villain, father, doctor, engineer, but also any words, moral or functional, for *patient,* and words for the motivational properties or agents, such as *drives, instincts, states of mind.* We may also have collective words for agent, such as *nation, group,* the Freudian *super-ego,* Rousseau's *volonte generale,* the Fichtean *generalized I.* (p. 20)

Thus, agents are not necessarily individuals. They may include groups, agencies, organizations, businesses, and even the government. In fact, most technology development will involve some of these collective bodies as active participants that shape the course of development. A government, for example, can easily change the course of technology development by intervening with special policies. A case in point is the development of computer technology in China. Since its early days of development in the early 1980's, although its overall path has been upward, this technology has seen many ups and downs in various aspects due to the many, sometimes conflicting, policies made by the Chinese government in its various efforts to promote certain aspects of the technology while suppressing others.

This relationship between technology development and the participants can be seen more clearly through Burke's agent-act and act-agent ratios. According to the agent-act ratio, the particular traits of the agent determine that only acts of a particular nature will be performed. When translated into technology development, this means that the particular nature of the participants—the elite or the masses, political or technical, receptive or resistant, government or individuals—determines to a large extent the fate of that technology. In a reverse perspective, that of the act-agent ratio, the nature of the act presupposes the kind of agent that performs it. In technology development, this means that if, for example, a technology fails in a particular context, although there can be many factors other than the participants that contribute to the failure, the participants must be of a certain nature that prevents the technology from being successfully developed—either they are too resistant to the technology, or they might be cognitively challenged, or they might have too many conflicting interests. In a different scenario, if a technology ends up being primarily a control tool, a symbol of power, the major participants, the agents, must have been a politically-driven group, possibly the government.

Act—Meaning Construction/Knowledge Creation

Act, according to Burke (1952), is "what was done," "what took place, in thought or deed" (p. x). Obviously, in technology development, act is the development of the technology. This seems fine when we take the term "technology" at face value. For example, in the case of printing technology development, the act is the invention of printing methods and tools. However, problems arise when we ask the question: what is technology? As I have shown in Chapter 1, technology refers to more than just the physical artifact; it carries cultural values, something called technical rationality. Thus we can't really say that the act is the development of that physical artifact. Since, as I have argued, what technology is really depends on how people define it, it in essence is only but a symbolic representation in the agent's mind. Technology thus becomes nothing more than the meaning constructed of it by the agent. Therefore, instead of the vague term of technology development, the act really refers to that of meaning construction. Since meaning when accepted by a community becomes knowledge, the act of meaning construction is in essence the act of knowledge creation.

Agency—Communication Medium

As the "means or instruments," the agent used to perform the act, agency describes how the act was done (Burke, 1952, p. x). The act-agency ratio tells us that the nature of the act determines that the agent can only use certain kinds of means. Since the act of technology development is really an act of meaning construction and knowledge creation, the act essentially becomes an act of communication. The media for communication thus become the agency—the means of performing the act. In this agency, the primary medium is no doubt language. At the same time, we also have other types of media, such as television, radio, newspaper, books, electronic mail, the Internet, etc. Which media get used depends on the context.

Purpose—Knowledge Access/Control

Any act has a purpose, which is what separates action from mere motion, which in turn distinguishes humanity from animality (Burke,

1966, p. 2-24). The big *why* behind technology development is obviously to have that technology, to make it serve us human beings. If technology development is essentially an act of meaning construction and knowledge creation, the purpose then is the possession of that knowledge, its access and control.

One of Burke's ratios is of particular relevance here, the purpose-act ratio. This ratio dictates that a particular purpose determines the nature of the act. If a particular group of participants, such as the political elite, wants to create a technology as a tool for control, they will construct the meaning of that technology in such a way that it serves the interests of the ruling class only and serves to strengthen their status of power in the social and political hierarchy. By the same token, a technical elite, with a purpose to confirm their authority in intellect, might so construct the meaning of that same technology that it represents a symbol of knowledge. Whatever the motive, the purpose of any participant groups in technology development would be to gain access to and control of the knowledge.

Attitude—Ideology

So, where does *attitude* come from? *Attitude* is the sixth term that Burke would have liked to add to the Pentad: "Many times on later occasions I have regretted that I had not turned the pentad into a hexad, with 'attitude' as the sixth term" (1972, p. 23). What is attitude and where does it belong? Burke (1952) answers the question this way:

> Where would attitude fall within our pattern? Often it is the *preparation* for an act, which would make it a kind of symbolic act, or incipient act. But in its character as a state of *mind* that may or may not lead to an act, it is quite clearly to be classed under the head of *agent*. (p. 20)

To Burke, attitude means two things: a preparation for the act or an incipient act. As an incipient act, it can also mean two things: "the *substitute* for an act" or "the *first step towards* an act" (p. 236, emphasis original). Either way, "an attitude is a state of emotion, or a moment of stasis, in which an act is arrested, summed up, made permanent and total" (p. 476). Clearly, such a definition would place attitude under the category of agent, which would render this separate section unnecessary.

There are two reasons why I am treating attitude separately. For one, if we deconstruct the Pentad, all the other four elements in the Pentad can be claimed to be more or less part of agent. The ability to act is an innate quality of the human agent; the purpose as motive is also an inherent part of the human being; even agency and scene are part of humanity as they make the human act complete. However, all these elements have their distinctive qualities that merit separate treatments from agent. The same is true with attitude. Although it is as much a part of agent as anything else, there is one quality that deserves special attention, which is my next, second point.

This second reason for a separate treatment of attitude lies in its social nature. Attitude is not an exclusively individual feature; it is in fact as much social as it is individual. As Burke (1952) puts it:

> A social relation is established between the individual and external things or other people, since the individual learns to anticipate their attitudes toward him. He thus, to a degree, becomes aware of *himself* in terms of *them*. . . . And his attitudes, being shaped by their attitudes as reflected in him, modify his ways of action. Hence, in proportion as he widens his social relations with persons and things outside him, in learning how to anticipate their attitudes he builds within himself a more complex set of attitudes, thoroughly social. This complexity of social attitudes comprises the "self" (thus complexly erected atop the purely biological motives, and in particular modified by the formative effects of language, or "vocal gesture," which invites the individual to form himself in keeping with its social directives). (p. 237)

Defining attitude as social thus invites a question: what is the source of individual attitudes? If attitude is a combination of the individual and the social, or, to borrow Freud's terms, the *id* and the *superego* what part of the superego shapes an individual's formation of the state of mind, the symbolic act or the incipient act, on an object or issue? This is where ideology comes in. Ideology represents a particular way of group thinking, a collective state of mind, within a particular culture. A cultural attitude is in essence the ideology of that culture.

The existence of ideology and its influence on individual attitudes and actions, according to Burke (1989), are inevitable:

> An "ideology" is like a god coming down to earth, where it will inhabit a place pervaded by its presence. An "ideology" is like a spirit taking up its abode in a body: it makes that body hop around in certain ways, and that same body would have hopped around in different ways had a different ideology happened to inhabit it. (p. 59)

This ideology serves as the driving force within human motive behind attitude, which directs the action of the human agent. Often times, it is not readily visible on the surface for us to detect it; rather, it often serves as the hidden cause for a particular course of action.

Why then do I choose to use ideology instead of attitude as an element in my rhetorical model of analysis? The reason is twofold. For one, individual attitudes, although social in nature, are temporal rather than constant. An individual's attitude does not last forever; it changes constantly when the context changes, even though it might be shaped by the same ideology. Ideology, on the other hand, is more stable. Often a particular ideology will last for an extended period of time before it gives way to a different ideology. A second reason is that different people, even if influenced by the same ideology, may have different attitudes. This is due to the differences in their individual experiences and contexts. Therefore, ideology is a much more reliable variable than its counterpart, attitude, which is the surface representation of ideology.

A Rhetorical Model of Analysis

Burke's pentad—the categorization of a rhetorical situation into five aspects: act, agent, agency, scene, and purpose (1943, 1969)—plus the sixth element, attitude, is thus a logical and good way to capture and depict the essence of the elements of the process of technology transfer and development. The six elements of my rhetorical model of analysis—exigency, ideology, participants, knowledge creation, knowledge access and control, and communication media—provide a convenient means to examine technology transfer and development as rhetorical acts of knowledge reconstruction and communication. These six elements comprise a rather dynamic picture of the process of technology transfer and development as dictated by the cultural, political, or economic circumstances. Ideology identifies the prevailing world view(s) governing people's attitude, and thus action, of a certain historical period. Participants are the ultimate agents that carry out the process. Knowledge creation deserves a special note because how that technology is represented rhetorically, i. e., what knowledge gets to be communicated to represent that technology, determines to a large extent the outcome of technology development or transfer. A related issue is the control over this knowledge. Control (power /authority) sheds light on those factors that influence or even dominate people's perception of the technology—what it is, how it should be used, who can use it, etc. Finally, communication media specifies the particular patterns of language, writing, and/or communication that affect the rhetorical process of technology transfer and development.

In the following section, I will discuss these elements in terms of both their generic nature and their unique representation in the Chinese context.

Exigency—The Initiating Factor of Technology Development

As I have shown, by exigency, I mean a certain compelling need that arises out of social, political, or economical circumstances, or a combination of these, which acts in a given context as an initiating factor that either leads to the invention of a particular technology or spurs its development. This exigency can be economically driven, such as cost effectiveness or efficiency; or ideologically driven, such as self-reliance; or politically driven, such as international preeminence. Identifying and understanding this exigency is a prerequisite to understanding the other elements that constitute the development or transfer process. It is the central piece of the puzzle that connects all the other pieces. In this sense, Jennifer Slack's (1984) explication of the structural causality between technology and society sheds much light on how a particular social structure dictates the emergence of technology:

> . . . technologies are not autonomous objects isolated from the social structure within which they exist. Rather, technologies are semiautonomous elements located within a specific historical configuration within a mode of production. Although a technology is thus a real object (a real machine or structure), it is at the

same time part of a changing whole that is structured in dominance. A technology, then, has no essence, and it therefore cannot reflect or express an essence, as is maintained in the expressive causal position. (p. 89–90)

Here, by saying that technology has no essence, Slack is obviously referring to the kind of technology that is bereft of cultural identity/ meaning, that is without historical and social contexts, and that certainly doesn't exist. Therefore, one of the first things to do when we examine the formation of technology is "to locate technologies within a particular historically constituted social whole and to explain the relationships between the technologies and the whole as it changes" (p. 90). This "particular historically constituted social whole" certainly includes all the elements I mentioned above. Such social and historical contextualization, then, will help us determine, among other things, the particular exigency to which the technology is a response.

The development of new technology as a response to a certain exigency finds ready examples in the history of technology development both in the East and in the West. For example, in discussing the invention, development, and use of some communication technologies in the nineteenth century, such as the telephone, the telegraph, the typewriter, duplicating methods, and filing systems, Joanne Yates (1989) attributes the emergence of these technologies to the impact of a new managing philosophy: that of systematic management, which replaced the old ad hoc, familybased management method. A direct result of this new management philosophy is the adoption of a system of internal communication, which evolved as a response to a new cultural value: efficiency. As part, and an important part at that, of this system of internal communication, the communication technologies mentioned above emerged, to a great extent, as a response to the increasing demand for efficiency. Therefore, efficiency was the apparent exigency element in the historical and social context of the development of those communication technologies in America in the nineteenth century.

The necessity of the existence of a compelling need, or exigency, as an initiating factor in technology development is also illustrated by the case of the rather late boom of press copying technology, which saw a lapse of almost a century between its invention and its wide use. "First patented by James Watt in 1780," press copying did not come into general and wide use until the second half of the nineteenth century (Yates, 1989, p. 26). That press copying did not gain immediate popularity at a time when hand copying was the only available method of duplication is seemingly bewildering. However, an examination of its social and historical context reveals this as a no-surprise phenomenon. According to Yates, the "delayed acceptance . . . stemmed both from early inadequacies of the technology itself and *from the absence of a compelling need in most companies of the earlier period*" (p. 27, emphasis mine). This compelling need did not come until early to mid-nineteenth century when systematic management dictated efficiency as the exigency of the time. Only then did this technology find its nourishing soil.

The phasing out of press copying technology several decades later, however, is yet another example of how cultural values (and the exigencies they dictate) impact and even determine the development path of technology. By the end of the nineteenth century, two other copying technologies became available: roller copying and carbon copying. To decide on the desirability of these copying methods, two government studies were conducted, the Keep Commission study of 1906 and the Taft Commission of 1910–1913. Four major criteria were used for evaluation: *permanency, authenticity, economy, and adaptability*, of which the first two were debated and the last two played a central role. Carbon copying prevailed.

What was not mentioned, as a fifth factor, in the government studies, was that "carbon copying was better suited to the newly developing internal communication" for its ability to produce multiple copies at one time instead of a maximum of only two copies that press copying was capable of (Yates, 1989, p. 45–50). What is noteworthy here is that economy and adaptability were valued while permanency and authenticity were devalued. Why? The answer again lies in the exigency of the time. Therefore, when several alternative technologies are available for the same purpose, the fact that only one or some but not the others prevail is the best evidence that exigency, be it economical, ideological, or political, plays a significant role in shaping the adoption, the use, and the development of technology.

Ideology—The Invisible Driving Force behind Technology

If exigency initiates technology development, ideology specifies the position of technology within its "particular historically constituted social whole" and designates the relationship between the part (technology) and the whole (the social context). In particular, ideology defines the relationship between individuals and a social phenomenon, and in our case, between participants and the technology. As Slack (1984) has put it, "due to this unique role of ideology, as both designating real relations but not necessarily revealing their essences, ideology is often referred to as the 'imaginary relation' between individuals and their real conditions of existence" (p. 85). The unrevealing nature of the designation is what makes ideology a sometimes seemingly invisible player in technology development and transfer. This invisibility no doubt adds to our difficulty in understanding why certain technology develops the way it does. It is, therefore, of particular importance, and also the purpose of this section, to define ideology in the context of technology development and to provide a perspective on the reigning ideologies throughout the history of China that have exerted impact on technology development.

Since technology development inevitably involves the participation of human agents, ideology works its impact by constructing, at least in part, the subjectivities within human agents. The way ideology works, as Jane Gains (1991) sees it, is as follows:

> As Althusser points out, ideology works through us, often with our own enthusiastic cooperation. In working through us, it "fits up" ideal subjects for the kind of society that must be produced in order to reproduce the capitalist mode of production. Ideology "recruits," as Althusser puts it, but it doesn't coerce subjects. In a way, it doesn't need to coerce because it is we who self-identify ourselves as these very subjects; we are the ones who see no other way of being because we recognize ourselves in the absolute obviousness of the existence that ideology constructs. (p. 27)

Althusser's confidence in and recognition of the role of human agents in the making of history is certainly questionable here. However, his recognition of the role of ideology in constructing subjectivities, and consequently culture, is a helpful concept that can aid us in our understanding of the relationship between ideology and technology.

Slack's (1984) interpretation of Althusser's concept of ideology sheds much light on our subject:

> *Ideology* is, stated at the most rudimentary level, "a system (with its own logic and rigor) of representations (images, myths, ideals or concepts, depending on the case) endowed with a historical existence and a role within a given society" (Althusser, 1970, p. 231). Ideology is not false consciousness, and its presence is not limited to the capitalist mode of production. Rather, ideology is a "structure essential to the historical life of societies," and *"an organic part of every social totality"* (p. 232). The way we live our lives, the way we live the relationship between ourselves and the world, the very conception of what "we" means is ideology. Ideology "naturalizes" its apparent reality; that is, it is a system of representations within which we see ourselves or live our lives in relation to our conditions of existence. Ideology does "designate a set of existing relations," but "it does not provide us with a means of knowing them. In a particular (ideological) mode, it designates some existents, but it does not give us their essences" (p. 223). (p. 85)

This at once real and imaginary relation between "individuals and their real conditions of existence," between participants and technology, is what makes the study of the role of ideology in every technology development and transfer all the more important. In the reciprocal relationship between ideology and social existence, which includes technology, ideology "produces subjects; that is, it defines and shapes for us our own subjectivity. It actively constitutes our experience of ourselves and thereby our conditions of existence" (Slack, 1984, p. 85). Such a structural causal position, in Slack's own terms, "posits that causes do not exist apart from their effects and that structure, which exercises effectivity, consists of its effects. . . . [It] posits technologies as very much a part of the social structure within which they arise and exercise effectivity" (p. 82). The significance of such a conception lies in that such a model of the relationship between technology and society "should provide us with the tools for intervening both before and after the appearance of technologies" and that "it should give us some way for determining specific locations for interventions such that our interventions will make a difference" (Slack, 1984, p. 82).

In his analysis of the impact of cultural values on technology, Feenberg (1991) offers a dialectic approach toward the relationship between ideology and technology, which, as he conceives it, is one of both resistance and compliance. "Ideological resistance [brings] technology into structural compliance with emerging cultural values," he asserts. "Such a transformation of values into socio-cultural facts occurs through the realization of these values at three levels: in ideology, particularly as embodied in laws; in economic realities and the interests of which individuals are conscious; and in the technological underpinnings of the social order" (p. 130). Whether Feenberg is implying that ideology is the highest of the three levels is not clear here, but the role of ideology as dictating the position of technology within the "particular historically constituted social whole" and as designating the relationship between technology and society /culture, and between technology and individuals, is overtly suggested, if not expressly stated.

Yates's (1989) analysis of the Du Pont case provides a good example of how management's ideological and philosophical orientation determines to a large extent the path of technology development and use within the corporate culture. As mentioned earlier, around the mid-nineteenth century, driven by the exigency of efficiency and the systematic management philosophy, most American businesses were adopting new communication technologies. Du Pont's management, however, seemed to be committed to a more traditional, conservative ideology, which greatly deterred the use of these new technologies in the company. Du Pont's first century (1802–1902) was marked by a strong conservativism in the family and the firm. This conservativism was also found in Du Pont's reluctance to adopt and use the new technologies. For example, Henry du Pont, head of the company from 1850 to 1889, insisted on using quill pens instead of steel pens, and hand copying instead of press copying. The younger generation at Du Pont was more receptive to new technologies. So, toward the end of the nineteenth century, there was some kind of a balance between acceptance of and resistance to technologies within Du Pont, resulting in some technologies being adopted while others were rejected.

This Du Pont dilemma reflected a clash between different ideologies, between the internal culture that sticks to an old ideology and the external culture that is more willing to embrace a new ideology. While the external culture of that period was characterized by the philosophy of systematic management, and thus an emphasis on technology to help achieve this efficiency, the internal culture at Du Pont was predominantly conservative, although the influence of the external culture was inevitably forcing its way in, especially through the younger generation at Du Pont. Such an ambivalent, if not hostile, attitude toward technology could be attributed to the management philosophy of Du Pont, which was more ad hoc than systematic at that time. Therefore, when systematic management took over at Du Pont in 1902, radical changes in the adoption and use of technology took place.

Although the Du Pont case illustrates the impact of ideological orientation on the use of technology within a corporate culture, its application to national and ethnic cultures is equally valid for reasons that defy explication. My discussion in the first chapter of the dominant ideologies in the history of cultural thought in China serves, in part, as a rationale for my subsequent chapters, where I discuss the significant cases of writing technology development in China and where such cases serve as illustrative examples of how ideologies (and various other rhetorical elements discussed in this chapter) have been at work in writing technology development.

Participants—The Activating Agent

The role of participants in my rhetorical model of technology transfer and development is obvious if we return, just for a brief moment, to the definition of technology transfer I offered toward the beginning of this chapter. As I have stated, technology transfer (and development) is a process of individual and collective construction and reconstruction of perspectives, values, language, knowledge, etc., which are conveyed through rhetorical means. With technology transfer and development being essentially a cultural event, participants as performers and carriers of cultural transformation hold the central position in this process. The significance of this is reflected in the fact that participants are the creators and determiners of exigency, the captivators and captives of ideology, the users of language and performers of rhetorical acts, the masters and slaves of power and control, and the creators and carriers of knowledge. Such a role places the participant at the center point of the technology transfer and development process, where all the elements of the process converge and where all interactions occur.

In his analysis of several different models of technology transfer and development, Doheny-Farina (1992) observes that these models acknowledge, though to varying degrees and on different theoretical underpinnings, the active role of the participant in this process. The information transfer model recognizes the role of the participant though it assumes, in an overly simplistic way, that everybody speaks the same language, has similar goals, comes from similar disciplines and uses similar technologies, that information is something to be transferred among participants that share similar worldviews (p. 12–13). The information gatekeeping model assigns this role to select individuals and assumes the existence of an information gatekeeper who translates the content of information and an information officer who directs to sources, with both being "active agents that shape information as they learn it and present it to others by conceiving of the world of information that is appropriate for their colleagues and by selecting and interpreting the 'information' that they see in that world" (p. 15). The informal information model goes one step further by maintaining that "information is transferred from a source to a destination but . . . information is 'predigested,' that is, interpreted and negotiated among participants in communicative activities" (p. 16). The information interface model, on the other hand, argues that "communication establishes an interface between groups . . . The interface is the relationship among individuals, not a channel through which information is acquired and disseminated" (p. 16). Examined closely, these models of technology transfer and development exhibit obvious inadequacies in their theoretical perspectives. It should be noted that my purpose of bringing up these models is not to recommend that we adopt their perspectives. Rather, they serve to illustrate that in our quest to achieve a thorough understanding of the technology transfer and development process, we have always acknowledged the role of human agents in this process, regardless of the degree of importance assigned to such a role.

In this sense, Williams and Gibson (1990) have offered a better alternative in their discussion of the communication-based model of technology transfer. They argue that "successful technology transfer is an ongoing interactive process where individuals exchange ideas simultaneously and continuously. Feedback is so pervasive that the participants in the transfer process can be viewed as 'transceivers,' thereby blurring the distinction between the source(s) and destination(s)" (p. 15–16). They further argue, "The technology to be transferred often is not a fully formed idea and has no definitive meaning or value;

meaning is in the minds of the participants. Researchers, developers, or users are likely to have different perceptions about the technology, which affects how they interpret the information. As a result, technology transfer is often a chaotic disorderly process involving groups and individuals who may hold different views about the value and potential use of the technology" (p. 16).

Doheny-Farina (1992) acknowledges the value of such a perspective. At the same time, however, he points out the problem with this view, noting that "they have separated facets of the rhetorical nature of technology transfer from communication" and that in so doing "they return to the old concept of information transfer" (p. 23). As an alternative, he offers a more complex view: "From the sources to the users, technological innovations are part of complex social, organizational, institutional interactions, interpretations, and negotiations. And every participant along the way constructs and reconstructs the innovation based upon his or her experience and worldview" (p. 7). According to such a perspective, the end result of the transfer process is different realizations of the original innovation/technology produced by the new cultural agents. Such a perspective recognizes not only the role of every individual participant in the process, but also the interaction among the participants and the interaction between participants and the social context. In other words, it is a more contextualized, more rhetorical construction of the participant role in the process of technology development and transfer.

Knowledge Creation—The Essence of Technology Transfer and Development

As a series of individual and collective constructions, reconstructions, and negotiations of meaning, the process of technology development and transfer is, in essence, a process of knowledge construction. As Doheny-Farina (1992) has argued,

> *At their core* these [technology transfer] processes involve individuals and groups *negotiating* their visions of technologies and applications, markets and users in what they all hope is a common enterprise. This means that the reality of a transfer does not exist apart from the perceptions of the participants. Instead, the reality—what the transfer means to the participants—

> is the result of continual conceptualizing, negotiating, and reconceptualizing. (p. 4, emphasis original)

The reality, which is constructed and negotiated by the participants, refers to the constructed meaning of technology, in other words, the knowledge about technology. Since people, as the enculturated agents, are the constructors of this knowledge, what is at play in the construction is individuals' experiences and worldviews and the culture's defining characteristics, which define the patterns of knowledge creation.

Such a knowledge construction occurs at two different levels, one individual and the other collective. At the individual level is the construction of meaning by several different groups of participants, including the administrators and regulators, the so-called political elites; the technical experts, the so-called knowledge elites; and the general public as the users. Each of these participant groups contribute their own interpretations based on their group identity and interests, and each of the participants within each group attributes meanings or interpretations to the technology based on their own individual experiences. Often, the political elites map out the knowledge at the ideological level while the knowledge elites construct it at the theoretical level. The general public, on the other hand, does not enjoy, although they are entitled to it in every way, the privilege of constructing the ideological and theoretical frameworks for the knowledge. However, since ideological and theoretical constructions are based, at least partly, on public perceptions, individual users do contribute to and shape these constructions in a way that is both fragmented and unified. It is fragmented in the sense that the general public contributes to the knowledge construction through individual efforts, and it is unified in the sense that such individual constructions are what constitute the basis of a collective cultural perspective. Because of the differences in experiences and perceptions among individuals and groups, the realities about the technology that they create are thus, more often than not, at odds with one another. These variously constructed realities are then mediated and negotiated into a common reality, accepted, more or less, by all parties involved, by way of the social structure that regulates the groups and individuals.

This knowledge construction is not complete until it moves to the collective level, where it acquires a cultural ethos. I have mentioned earlier that technology is comprised of two elements, the concrete tools and the technical rationale. The technical rationale of a particular technology varies from country to country, from culture to culture, and is heavily embedded with the values of a particular culture. Therefore, the cultural ethos of technology refers to its value-laden, culture-specific identity. Fleron (1977c) defines the national cultural ethos as "the historical and cultural traditions" of a country; "it includes religious factors; the patterns of authority structures; the level of economic development; the system of social stratification; and certain traditions in science, philosophy, and the arts" (p. 3).

So, how is culture knowledge? How is knowledge cultural? And how is technological knowledge cultural knowledge?

First of all, knowledge and culture overlap and are sometimes equated. For example, in his discussion of Lenin's concept of culture, Feenberg (1977) defines culture as civilization, as knowledge, and as ideology. Culture as civilization is "the material and spiritual heritage of a society, the ultimate measure of its level of development;" culture as knowledge means "the general culture and skills required to participate intelligently in the economic and social life of a modern nation;" and culture as ideology refers to "strictly class-bound world-views in competition for the mind of the masses" (cited in Fleron, 1977a, p. 469). We can take this concept one step further and argue that civilization and ideology are part of what makes up knowledge. In this sense, any knowledge will bear some cultural imprint.

Second, since we have defined earlier that technology development and transfer is in essence a process of knowledge construction, it is not hard to see how this knowledge is cultural, given the reasoning in the last paragraph. As William Dunn (1977) sees it, "technical rationality, technology, and technique are conceptualized and defined as elements of culture, and culture itself is taken as constitutive of knowledge, belief, art, morals, laws, customs, habits, traditions, and symbolic capabilities" (p. 364). Constructing knowledge about technology at the collective level means imprinting the cultural ethos on the knowledge constructed at the individual level. Mao Ze Dong, the former leader of China, once provided a vivid interpretation of the cultural tinting of this meaning and knowledge construction:

> China has suffered a great deal from the mechanical absorption of foreign material.... We should assimilate whatever is useful to us today not only from the present-day socialist and new-democratic cultures but also from the earlier cultures of other nations.... We

> should not gulp any of this foreign material down uncritically, but must treat it as we do our food—first chewing it, then submitting it to the workings of the stomach and intestines with their juices and secretions, and separating it into nutriment to be absorbed and waste matter to be discarded. (Dernberger, 1977, p. 225–226)

An interpretation of this metaphor provides a good summary of the process of transfer and the construction of knowledge about technology. First, the technology (the food) is brought into contact with a new culture (gulped). Then it is studied and understood (chewed) before it is adapted and transformed by the new culture (submitted to the workings of the stomach and intestines with their juices and secretions). Finally, a reality, or knowledge, is constructed by the participants which redefines the technology by creating a new technical rationality based on the workings of the new culture (the nutriment to be absorbed) to replace the old, alien technical rationality (waste matter to be discarded). Examined from such a perspective, knowledge about technology is thus inevitably instilled with a cultural ethos due to the way it is constructed.

Knowledge Access and Control—Powers and Privileges

A discussion of knowledge construction entails an examination of the issue of control, power, or authority, which shapes and determines a number of aspects in the knowledge construction process: who decides what gets constructed about the technology, who decides what knowledge gets disseminated, and who decides how that knowledge is disseminated. Since there are different participant groups in the knowledge construction process and they each create their own versions of realities/meanings/ knowledge about the technology, discrepancies in their perspectives about the technology are inevitable. These discrepancies, though the source of conflict in knowledge construction, do not necessarily lead to irreconciliation. This is mainly due to the fact that the social structure which regulates the society and its various elements dictates, in much the same way as it does in other areas, a hierarchical structure that governs knowledge construction. As a result, knowledge construction is not a purely egalitarian or democratic process, though, theoretically, it should be.

The undeniable fact is that possession of and access to knowledge are not equal among participants. The mere fact that elite groups exist, as mentioned earlier, is itself an indication of inequality in status between various participant groups in technology transfer and development with regard to access to knowledge, which includes its construction and distribution. Fleron (1977b) seems rather perplexed by the cause of the existence of a political elite and knowledge elite (what he calls "a technical and/or managerial elite"). He does not make a distinction between these two elite groups and attributes their existence to either political choices or technological imperatives (p. 48). The difficulty in distinguishing between the mastery of knowledge and its control reflects the interrelationship between knowledge and power. Roger Garaudy (1970) has the following statement about the difficulty in isolating one from the other:

> The staggering increase in our technical mastery of nature entrusts to a few handfuls of men a fund of knowledge and organization that gives them a terrifying power. This technocratic cleavage between the managers and the masses is the law in monopolistic regimes in which the concentration of resources is a class matter. In fundamentally different conditions, the objective difficulties created by this situation raise problems for the development of a true socialist democracy. (cited in Fleron, 1977c, p. 48)

Although Fleron and Garaudy's main concern is how the existence of this technical and/or managerial elite affects the "the development of a true socialist democracy," their discussions do point to the existence of inequality among participant groups in terms of knowledge control. My concern here is how this unbalanced allocation of power affects the process of knowledge construction and its dissemination.

As stated earlier, participants in technology transfer and development can be categorized roughly into three groups: the political elite, the technical elite, and the ordinary users. If we place these three groups on a continuum of degree of control over knowledge construction and dissemination, we would find the political elite at one extreme with the most power and the ordinary users at the other, with the technical elite situated somewhere in the middle. The political elite is comprised of policy makers and interpreters at various levels, from high-level officials who make policies that regulate the field of technology to middle-level managers who interpret policies. This group maps out the path of technology development at the ideological level by outlining technology policies that are in keeping with the state political ideology and the long-term and short-term political and economic goals of the government. It is not hard to see then that government plays a pivotal role as a shaping force of the ideological proclivities of the political elite. The knowledge constructed by this group about technology thus carries a strong mark of political inclination, especially in socialist regimes, where democratic thinking is limited to a rather controlled level.

At the other end of this continuum, where there seems to reside the least power and control over knowledge, we find the ordinary user group, which is comprised of unskilled workers and the general public. Under the capitalist regime, Marx sees this group as being in a position of cultural incapacity and of an "inability to understand and master production on the basis of their narrow experience of it," arguing that "capital establishes its reign over labor through a division of labor that disqualifies the worker and renders the worker helpless before the massed forces of knowledge embodied in capital" (Feenberg, 1977, p. 80). Presumably according to Marx, under the socialist system, such a problem would not exist due to the fact that class distinctions do not exist in socialist countries. First, "under socialism, . . . no fundamental conflict of interest would divide organizers and organized into a ruling and a ruled class;" second, "under socialism, workers would control not only day-to-day production, but also the long-term reproduction of society" (Feenberg, 1977, p. 89). However, any close examination of socialist regimes would reveal that the issue of knowledge control in socialist countries is no less complicated, which I show as follows.

There are two major flaws to Marx's argument of class distinction. One is that the kind of socialism where no class distinction exists according to his rather romantic depiction is the ideal socialist society, which, unfortunately, does not currently exist. The other flaw lies in his failure to observe the multiplicity in the ways of categorizing of classes. Marx's definition of class distinction is solely based on economic status (i.e., the individual's accumulation of capital and wealth). The other criteria sets that are often intuitively used by the public, but not so well elaborated on and often ignored by scholars, to distinguish classes are political prominence and intellectual wealth. It is these two ways of categorizing that apply to socialist societies and that precisely relate to my central issue of knowledge and power here.

If we look at class distinction in terms of political prominence, we have two groups: the political elite, which is politically powerful in deciding what and how technology gets used, and the general public, which does not have a prominent role in the construction of this knowledge. If we use intellectual wealth as a divider, we have the technical elite, who are intellectually empowered by their wealth of knowledge, and the general public, which does not possess the kind of technical knowledge that would put them in a more egalitarian, if not more privileged, position in determining what realities should be constructed about a particular technology.

If, according to Marx, no class distinctions exist in a socialist country, then no conflict of fundamental interests would occur, and the construction of knowledge about technology would be greatly simplified. However, in a socialist country, while the government looks at technological development in terms of generalized and universalized interest/motivation for its working class, individuals perceive it more in terms of the concrete, contextualized, and in Feenberg's (1977) term, "empirically identifiable" interests. Unfortunately, there is almost always a gap between the two perceptions in that while the former aims at reinforcing its control over the people, the latter has as its goal self actualization through "the development of human attributes and capacities . . . mediated by material goods" (p. 93). Put in simpler terms, while the government aims to consolidate its rule over the people by controlling their access to knowledge, the people seek to better themselves by means of increasing their knowledge. Unfortunately, such a conflict of interests is not a matter to be easily explained away by a mere denial of class distinctions.

In contrast, the position of the technical elite constitutes a more intricate case. Posited somewhere near the middle on the power continuum, this group at once feels a kind of subordination as the nonelite worker group does and possesses some of the power that the political elite enjoys, though in a different respect. On the one hand, this group does not have the power to map the ideological path of technology development and hence its knowledge construction. On the other hand, they are endowed with the power to interpret, from a technical standpoint, what technology means and how it should be perceived. This power, enabled by the technical knowledge that they possess, is at once a blessing and a curse to the technical elite. It is a blessing in the sense that as a missing link between the political elite and the general public, the technical elite is looked upon by both other groups as not only a contributor but, more importantly, as a negotiator in the process of meaning construction. Without the technical elite, knowledge construction about technology is undoubtedly impossible. However, this power is also a negative, in that the political elite, being aware of the threat to their position and the possibility of being replaced in the power structure by the technical elite, constantly attempts, by all means, to place them in a subordinate position. On the other hand, the general public looks at them as more or less an elite group and considers their position as much more privileged than their own. As a result, the technical elite may find themselves in an awkward position, one in which they don't belong to and are not accepted by either group. Moreover, being in such an ill-defined position as they are, and compelled to solidify their positions of restrained power, the technical elite must tend to their own special interests, which obviously are not totally in keeping with those of the other two groups.

Does this conflict of interests due to the unequal distribution of power between the three participant groups mean that knowledge construction of technology will always lead to a hopeless situation of irreconciliation?

Obviously not. The interrelated positioning of these three participant groups in the power structure of social formation leads to mutual containment and a resulting delicate balance, which, more often than not, results in a negotiated consensus between these groups about what constitutes the meaning of a particular technology, though such a consensus is built on the basis of conflicts. In his discussion of the relationship between technology, values, and political power in the Soviet Union in the face of a computer revolution, Hoffman (1977) provides a unique perspective on the dual role of experts, who are both empowered and threatened by the development in information technology. Table 1 reflects the ambivalent role of experts examined from two different points of view. What is worthy of note here is that the simple fact that such an ambivalent role is allowed to exist in this technical elite group, alone, seems to suggest, at least implicitly, that the inequality in power distribution in the process of knowledge construction can achieve a certain balance when molded in the structure of social formations. Constrained in such a social position, the technical elite group is unlikely to fundamentally change its status in the social structure, either to ascend to a position of sheer and absolute control over knowledge construction or to descend to a position completely deprived of intellectual and political power.

Table 1. Experts' Role Examined from Two Perspectives (Hoffman, 1977)

Knowledge as an instrument of political power in advanced industrial systems	Experts' primary social role: to legitimize policy decisions made by the real holders of power.
Societal complexity and rapid rates of change have the effect of making existing forms of knowledge and information obsolete. This increases the demand for new knowledge and information.	Major policy alternatives pose value choices that are inherently in conflict with each other.
The problems of advanced industrial society require specialized knowledge and information. This establishes the primary social role of experts.	Value disparities reflect the balance of power within the political system.
The problems of industrial society are increasingly amenable to solution through the application of existing knowledge and information.	Choice, therefore, symbolizes the 'victory' of a particular structure of power.
Because of the technical complexity of most policy decisions, experts are increasingly brought into the decision-making process to supply specialized information and advice.	After the decision is made, policymakers look for ways to legitimize their decisions. Technical explanations and justifications serve to diffuse conflict.
The political power of experts increases due to this social role. Politicians, because they are dependent on experts for knowledge and specialized information, witness an erosion of political power.	The image of the expert who is 'above politics' is a useful legitimizing tool. Moreover, since they are expendable, experts serve as convenient scapegoats for policies that have failed.

Along a similar line, we can also claim that neither is the political elite in absolute control of technology development and knowledge and cultural construction, nor is the general public completely powerless and void of impact in the shaping and constructing of the cultural ethos in and meaning of technology. For example, as Feenberg (1977) has argued, when the goals of technology development correspond to the development of socialism and "have the status of objective realities, sociopolitical and ultimately cultural constraints, the 'interests' of [the] workers may fundamentally influence technological development" (p. 104). Feenberg's ambivalent perspective might have allowed him a rather optimistic stance here in his assumptions about the impact of technology and about the conditions of democracy required for such a possibility. However, since the values and beliefs of the workers and the other sectors of the general public constitute the infrastructure of the culture, their experiences and perceptions do shape, in a fundamental way, the superstructure of the culture. Thus, instilling a cultural ethos in the knowledge construction about technology ultimately entails the inclusion of perceptions of technology by the general public.

On the other hand, building on Richard Baum's notions of "techno-logic" and "ideo-logic," Feenberg (1977) argues that, in the conflict between the technical elite and the political elite, which in essence is a struggle between "techno-logic," technological imperatives formulated by modernization theory, and "ideo-logic," socialist goals imposed upon the modernization process, techno-logic has a certain power that ideo-logic lacks. "Techno-logic is always presented as something 'real,' substantial, objective, almost spontaneous in character, like a natural process. Ideo-logic is a matter of human will and goals. it is 'voluntaristic'; that is, it lacks ultimate force in contact with techno-logic" (Feenberg, 1977, p. 101). Motivated by a technological deterministic perspective, Baum is certainly overstating the case, and I have no intention of endorsing such a stance. Nevertheless, it does point to the fact technology development has its due impact on the culture and its various elements and that the technical elite possess certain power derived from intellectual knowledge that the political elite precisely lack.

In conclusion, the discrepancies in the level of knowledge and power about technology between the three participant groups make it inevitable for them to each vie for their special interests and at the same time make it necessary for them to cooperate with one another so as to combine their strengths in order to attain a common good that will benefit all parties involved. The balance will be achieved one way or another. Nevertheless, it is unduly optimistic to presume the possibility and ease of such a balance, as Rodavan Richta, et al. have warned us:

> There is nothing to be gained by shutting our eyes to the fact that an acute problem of our age will be to close the profound cleavage in industrial civilization which, as Einstein realized with such alarm, places the fate of the defenseless mass in the hands of an educated elite, who wield the power of science and technology. Possibly this will be among the most complex undertakings facing socialism. With science and technology essential to the common good, circumstances place their advance primarily in the hands of the conscious, progressive agents of this movement—the professionals, scientists, technicians and organizers, and skilled workers. And even under socialism we may find tendencies to elitism, a monopoly of educational opportunities, exaggerated claims on higher living standards and the like; these groups forget that the emancipation of the part is always bound up with the emancipation of all. (cited in Fleron, 1977, p. 49–50).

If we have reason to be wary of the technical elite, we are equally justified to guard against the concentration of power in the hands of the political elite. We can only hope that, with due intervention by the general public, technology development and its knowledge construction will assume a course that is in the best interests of the society as a whole.

The Medium of Knowledge Distribution and Communication

I have spent much of this chapter arguing and proving that technology development and transfer is a rhetorical process of knowledge construction. If it is indeed a process of constructing and reconstructing ideas, how those ideas are communicated among participants and to the general public so that certain technology becomes technology—the real, enculturated, and accepted technology in a specific culture—

is, then, an issue of importance because the distribution and communication medium is one of the factors that affect the construction of meaning. What I mean by channels and media of distribution and communication includes the political and social mechanisms that are essential to the distribution of knowledge as well as popular media of public communication. There are two aspects entailed in this discussion of the distribution and communication of knowledge about technology. One is the issue of how this knowledge is communicated and distributed, namely, the channels and media of communication. The other has to do with the particular focus of this project. Since this is a discussion of the development of writing technologies, we cannot fail to consider how the particular language of a culture is used as a communication medium, which falls under the preceding issue, and how language characteristics, in particular, affect the development of certain specific writing technologies.

With regards to the channels and media for communicating and distributing technical knowledge, Feenberg (1977) proposes the following four mechanisms as conditions necessary for public access to technical knowledge in a socialist society, such as China. First, in a transitional society, there needs to be a reward structure that emphasizes "public models of consumption of goods and services and 'ideological' motivations for the acquisition and application of skills" (p. 97). This reward structure would be an important mechanism that would enable a shift in access privileges toward the masses. Second, the socialist, or any other transitional, society needs an organizational structure that will "insure the continuing and enlarged access of workers to the kinds of knowledge required to perpetuate and increase their power in society" (p. 97). Operationalized, this translates into a more egalitarian investment in the training of the workforce, a more evenly distributed cultural and technical qualification of the labor force, and, essentially, a more democratic means of distributing knowledge. Third, the transitional society needs to have a system in which "the exercise of authority by highly trained professional and managerial personnel" will contribute to "the enlargement of the workers' initiative and control throughout society" (p. 97). This would mean fair training and job definitions that "discourage technocratic and encourage democratic attitudes" and that promote democratic rules and roles concerning such things as "the procedures and criteria of innovation, the position of science in society, the restrictions on and context of the exercise of managerial authority, the taste and ideology of workers in mass communications and education, and so on" (p. 97). Fourth, the transitional society must make available "formal institutions . . . through which the workers can intervene in the activities of those with power and authority" (p. 98).

Granted that while such a perspective may seem a little optimistic, or even romantic, it does spell out, though in rather abstract terms, the social mechanisms required to enable the knowledge communication (i.e., knowledge construction) process. One important implication to be derived from this perspective is that when we examine knowledge distribution and communication, we cannot simply look at the communication media in an isolated manner. Instead, we must always take a contextualized approach to the process of knowledge construction so that we can recognize the true ways by which the medium of communication and distribution can affect and contribute to the creation of meaning. The rationale behind such a notion is that communication and distribution media, when susceptible to the influence of various factors such as ideological orientation, government control, public attitude, and other social mechanisms we mentioned earlier, can never be isolated from the context in which they are used and from the ideas being created and communicated because they directly affect the outcome of knowledge construction.

A case in point is the status of printing in the early days after its invention in fifteenth-century Europe, when it was not attributed as high a significance as it seemed to deserve. The main reason, according to Eisenstein, is that the historical context, with its particular social mechanisms and circumstances, did not facilitate a systematic construction of meaning/knowledge about printing. Instead, it allowed for a segmentation of the related aspects of printing, with the history of the book housed in the field of library studies, the history of printing in technology studies, and type, design, layout, and lettering in the school of design. As Elizabeth Eisenstein (1979) puts it, "when ideas are detached from the media used to transmit them, they are also cut off from the historical circumstances that shape them, and it becomes difficult to perceive the changing context within which they must be viewed. This point is not only pertinent to most histories of Western philosophy or literature; it also applies to most treatments of the history of science and of historiography" (p. 24).

A second aspect that deserves a special note for the purpose of this particular project is language, for two very sound reasons. One is that language, as an important medium of communication, cannot afford to be ignored. The other is that since my topic is mainly concerned with the development and transfer of writing technologies in China, the unique features of the Chinese language is a topic that has to be addressed because such features have affected, in fundamental ways, the development of some of the writing technologies in China, most notably that of the computer in the last couple of decades. The power of language to create knowledge is well documented by researchers (e.g., Ong, 1967, 1971; Burns, 1989; Havelock, 1986; Eisenstein, 1979). As Alfred Burns (1989) has argued, language and literacy are more than a means of communication. Burns identifies a historical correlation between literacy and intellectual advance, which viewpoint is shared by many researchers.

I will not go into further details on this issue in the present chapter. I will, however, address the issue when I discuss specific writing technologies in the subsequent chapters.

As I have shown, my rhetorical model of analysis posits that since technology transfer and development is essentially a rhetorical process of the construction, reconstruction, and negotiation of meaning, this process must be examined in terms of how different rhetorical elements function in the rhetorical situation of meaning construction. The six different elements, namely, exigency, ideology, participants, knowledge creation, knowledge control, and the medium of communication and distribution, contribute to the essential act of knowledge construction in the process of technology transfer and development. At the same time, a comprehensive perspective of any technology transfer and development entails a thorough examination not only of the workings of these individual elements but also of the interaction and interrelationship between these elements. In the next chapters, I will apply this rhetorical perspective to the examination of specific cases of writing technology development in the history of China.

4 Oracle and Bronze Inscriptions

> If we are to understand the China of today and tomorrow, we must know something about the China of yesterday and the day before yesterday. Chinese history is unlike the history of other countries. It is not a story, like that of Egypt, Greece, or Rome, of the early springing up, flowering, and dying out of a great civilization; nor is it like the story of Russia, which grew out of the Dark Ages into a modern nation in a few hundred years; even less is it like the story of the settling of a new land like America. China's history is the story of the oldest civilization in the world which has never died. (Lattimore, 1942, p. 4)

The above statement by Eleanor Lattimore about the history of China certainly also applies to the history of writing technology development in China. This enduring civilization has provided a rich and meaningful context for the development of various writing technologies as well as a wealth of resources for the examination of the six elements of the rhetorical process of technology development—exigency, ideology, participants, knowledge construction, knowledge access and control, and communication medium.

As has been argued in the previous chapters, any examination of writing technology development entails an exploration of the ways in which technical and social dimensions of the technology intersect to make that technology a concrete reality. As Feenberg (1991) puts it, "only at the point of intersection of technical and social determinations is this or that concrete technology identified in its specificity and clearly distinguished from among the wide range of possibilities supported by the available technical resources" (p. 130–131). A technology, as Feenberg has pointed out here, embodies both technical and social dimensions, with the latter probably being no less significant than the former in the formation of that technology. Certainly this is true of the history of writing technology development in China, as its history is one of fluctuations, with some periods nurturing more writing technologies than others. This warrants, to some extent, the contention that any writing technology development requires particular social conditions. Technical dimensions are the prerequisite of technology development, but they alone are obviously insufficient, accounting for the fact that different historical periods witnessed varying degrees of success in technology development. Instead, technical dimensions must intersect with the social dimensions in the right manner and at the right time to make technology development happen. Identifying this "point of intersection of technical and social determinations" for each of the major writing technologies in the history of China is the mission of this and the subsequent chapters.

Before I discuss the particular writing technologies in detail, an overview of the chronology of the major dynasties in the Chinese history and the major writing technology development would be helpful in contextualizing the particular writing technologies. Table 2 is but a sketchy chronology of the roughly five thousand years of recorded history of the Chinese civilization. These will be the focus of my discussion, although scientific archaeological data document the history to much earlier historical periods.

Table 2. Chronology of Main Chinese Dynasties

Dynasty	Dates	Writing Technology
Xia Dynasty	22nd–16th Centuries BCE	Oracle inscriptions
Shang Dynasty	16th Century BCE–1066 BCE	Oracle inscriptions Bronze inscriptions
Zhou Dynasty	1066 BCE–221 BCE	Bronze inscriptions Pens, ink, and early forms of paper
Qin Dynasty	221 BCE–206 BCE	
Han Dynasty	206 BCE–220 CE	Modern form of paper
Three Kingdoms	220 CE–280 CE	
Western Jin	265 CE–316 CE	

Eastern Jin	317 CE–420 CE	
Sixteen Kingdoms	304 CE–420 CE	
Southern and Northern Dynasty	420 CE–589 CE	
Sui Dynasty	589 CE–618 CE	
Tang Dynasty	618 CE–907 CE	Block printing
Five Dynasties	907 CE–960 CE	
Song Dynasty	960 CE–1279 CE	Movable type
Yuan Dynasty	1279 CE–1368 CE	
Ming Dynasty	1368 CE–1644 CE	
Qing Dynasty	1644 CE–1911 CE	
Republic of China	1911 CE–1949 CE	The Chinese typewriter
People's Republic of China	1949 CE–Present	The computer

Two disclaimers need to be made here. First, the dynasties included in the table represent the major dynasties during particular historical periods. Certain historical periods, however, witnessed the overlapping and/or co-existence of several dynasties as none of them were able to rule the entire territory of China. For example, in the span of the Song Dynasty (960 CE–1279 CE), there also existed the Liao, Western Xia, and the Jin dynasties, each of which spanned part of that period (see Fei, Xia, et al.). Second, although the writing technologies listed in the table are assigned to their respective historical periods based on their approximate date of origin, some of them spanned several historical periods, sometimes originating in one dynasty, then leveling off during its development, then being revived again in a much later period. This complexity in the fluctuating development makes our examination of the rhetorical context even the more intriguing.

Oracle Inscriptions and Their Uses

The exact date of origin for the Chinese language has always been a topic of contention and controversy. One popular argument considers the ceramic inscriptions of the Neolithic period as the earliest evidence of written Chinese characters (Shen, 1989, p. 428). These inscriptions on various pottery products were simple symbols rather than elaborate characters. Although many of these symbols are hard to decipher, they were obviously widely used, as they were found on various pottery relics unearthed from such culturally diverse sites as Yangshao Culture in Xian; Majiayao Culture in Qinghai; Dawenkou Culture and Longshan Culture in Shangdong, Hebei, and Henan; and Majiabang or Liangzhu Culture in Shanghai and Zhejiang. For example, twenty seven kinds of symbols were found on over one hundred pieces of pottery unearthed in Xian, while more than fifty kinds of symbols were identified on pottery pieces discovered in Qinghai (Shen 1989, p. 428–9). Most of these symbols are made up of simple strokes and are more or less the same in form. Usually each pottery piece carries only one symbol, either inscribed or drawn (*Common Facts*, 1980, p. 268–9; Shen 1989, p. 428–9). Archaeologist and writer Guo Moruo considers these ceramic inscriptions "symbols of a linguistic nature" and thus, perhaps, the origin of the Chinese language (Shen 1989, p. 429). Wang Li, a well-known Chinese linguist, even traced the origin of the Chinese language to more than ten thousand years ago (Yang, 2005).

Many researchers, however, contest such a claim, arguing that the inscriptions resembled more random, though maybe meaningful, symbols than systematic characters and that the archaeological evidence unearthed so far is too piecemeal and sporadic to warrant such a claim (Shen 1989, p. 428–9). In contrast, the oracle inscriptions that originated in the early Shang Dynasty (around the sixteenth century BCE) were far more systematic, had four or five thousand characters, were already equipped with means of character creation and use, and had rather systematic grammar and syntax; for this reason, oracle inscriptions were more widely accepted as the earliest systematic writing system of China.

Oracle inscriptions are characters inscribed by the ancients of the Shang Dynasty (sixteenth to eleventh century BCE) on tortoise shells and animal bones (Du, 1998). They were first discovered in Xiaotun in Anyang County, Henan Province, in 1899 (Xia, et al., 1979, p. 1673). Several digs at the site have unearthed more than 100,000 pieces of shells and bones with inscriptions. A total of around 4,500 characters have been identified, and about 1,700 deciphered (Xia, et al., p. 1673).

The tortoise shells and animal bones were originally used by the Shangs about 3,500 years ago as a tool for divination, and "many of

Oracle and Bronze Inscriptions

the inscriptions had to do with soothsayers and oracles" (Lattimore, 1942, p. 25). A very superstitious people, the Shangs would ask gods for revelation on occasions of or before a windstorm, thunderstorm, famine, epidemic, hunting, or war. Accounts about the methods of inscription among researchers are not exactly consistent, but they all seem to involve the use of fire. According to Rodzinski (1984), inscription was done via a method called "scapulimancy, the cracks produced by touching them with a heated object [, which] provided the answer to the question posed" to god about the outcome of the event (p. 19–20).

A different account (Du, 1998) has it that a hole was drilled on the interior of the tortoise shell, which was then put on a fire to see what cracks would form on the obverse side. The cracks were interpreted to predict the outcome of that event.

The content of the oracle inscriptions, according to Peng et al. (1989), can be classified into two main categories: divination and narration, with most falling into the former category. Divination inscriptions record the act of predictions about all aspects of people's lives, such as weather, gods, and harvests. Divination text seems to follow a particular organization: a preface, the topic, the prediction, and then the result. For example, one divination piece recorded that the King of Shang predicted a rainy evening that day followed by sunny weather the next day; the results record indicated it indeed rained that evening and indeed turned sunny the next day (Peng et al., 1989, p. 311–12).

Oracle inscriptions of the second category, narration, record people's observation of particular events. The main content includes the preparation and other details of the divination event preparation, such as the source, quantity, and inspector of tortoise shells; various charts, such as chronologies and family trees; and other, non-divination events, such as hunting and battles (Peng et al., 1989, p. 313–14). Most of these narrations are records of events related to the royal families of the Shang dynasty.

On one of the oracle bones unearthed from the Shang period, the left half reports a lunar eclipse people observed, and the right half records a nova. Both are dated about second millennium BCE The latter says in part: "On the 7th day of the month . . . a great new star appeared in company with Antares." Though by no means complex by modern standards, the message expressed by the characters is quite complete, suggesting that the repertoire of characters available during that period may already have been quite well developed. Many characters in oracle inscriptions are pictographs in their primitive forms. However, these pictographs are considered to have formed a rather systematic script: "a well-structured script with a complete system of written signs" (Du, 1998). A complete system of the Chinese script, according to Jian (1979), especially the modern script, is considered to have the following six features: (1) hieroglyphics, (2) self-explanatory characters, (3) associative compounds, (4) phonetic loan characters, (5) pictophonetics, and (6) mutually explanatory or synonymous characters (p. 27). The oracle inscriptions not only have a large number of characters but also possess all the six features (Jian, 1979, p. 27). If pictographs represent a primitive form of script, the presence of all these six features undoubtedly renders oracle inscriptions a systematic script. This systematic feature allows the oracle inscriptions to record rather detailed and accurate accounts of various events. For example, one large scapula from the reign of Wu Ding records several divination events that occurred over a period of at least 30 days in the fifth and sixth months. In each divination, the record of results confirmed the king's forecast of the disaster. The inscriptions, according to Keightley (1999), records:

> [Preface:] Crack-making on *guisi* (day 40), Que divined: "In the next ten days there will be no disasters." [*Prognostication:*] "There will be calamities; there may be someone bringing alarming news." [*Verification:*] When it came to the fifth day, *dingyou* [day 34], there really was someone bringing alarming news from the West. Guo of Zhi [a Shang general] reported and said: "The Tufang [an enemy country] have attacked in our eastern borders and have seized two settlements." The Gongfang (another enemy country) likewise invaded the fields of our Western borders." (p. 242)

These accounts, according to Lattimore (1946), present a more vivid and accurate picture of that historical period. They are more vivid in that the details in the inscriptions reveal more about "the way people thought in those times" than other written records, and they are more accurate in that, unlike other written records, which tend to exaggerate "great deeds to impress posterity," they provide more accurate information about the way people lived (p. 26). Though by modern standards these oracle bones are truly primitive, they nevertheless represent probably the first breakthrough in the development of writing technologies at the onset of the recorded history of the Chinese civilization.

Bronze Inscriptions and Their Uses

Around the same period (Shang and Zhou dynasties) when oracle inscriptions were widely used, a second form of written records—the bronze inscriptions—also emerged and prevailed. Bronze inscriptions were characters inscribed on bronze containers. The Shang Dynasty saw the invention and use of various bronze ware such as knives, axes, daggers, chisels, etc., used as both weapons and agricultural tools. However, bronze inscriptions were found almost exclusively on bronze containers only. Such containers, according to Peng et al. (1992), served mainly sacrificial, ritual, ceremonial, and gift-giving purposes.

The first main use of the inscriptions on bronze containers was as a symbol for a particular clan, something similar to the modern family logo. This use was featured mainly on early bronze containers, where inscriptions were usually short and simple, containing mostly one or two characters only. Such characters typically indicated the name (symbol) for the family.

A second use, also of early bronze inscriptions, was as a mark for the shrine of a particular ancestor. The Shangs were a particularly sacrificial people, holding frequent sacrificial and memorial rituals for their forefathers. During such rituals, bronze containers were used for holding offerings for the ancestors, and the inscriptions on the containers served to identify the particular ancestor the offers were for to avoid confusions. A later evolution of these first two uses of bronze inscriptions was to combine the clan name with the ancestor's name to indicate which ancestor the ritual was held for by which clan. These inscriptions were still rather short, typically containing four or five characters only (Peng et al. 1992, p. 335).

A third use for bronze inscriptions was more narrative than ritualistic, that is, they were used to record historical events. In the later Shang period, bronze ware making made some significant progress, and more bronze containers were made for inscription purposes. At the same time, the inscriptions on these bronze containers became longer and more elaborate. For example, the longest bronze inscription unearthed from the Shang period contained 47 characters and was made in memory of the owner's deceased mother. It recorded the owner's heroic event in which he acted as the guide in a battle led by the Shang Emperor, won honor, and was cited for his laudable actions (Peng et al. 1992, p. 336). By Western Zhou, the ensuing period after Shang, bronze inscriptions had evolved into very elaborate narratives. As Shaughnessy (1999b) notes, "the inscriptions run the gamut from the briefest mention that 'So-and-so makes the vessel' [in the Shang period] to narratives of several hundred characters [in Western Zhou], recording such events as appointments at court, victories in battle, and successful legal suits" (p. 296–297). For example, the DaYuDing, a famous bronze vessel from the Western Zhou period, contained 291 characters in its interior. It recorded the following admonitions by the Zhou Emperor to a person named Meng: the Yins were destroyed as a result of their indulgence in drinking while the Zhous prevailed because of their abstention from drinking; Meng should give his best efforts to aid the emperor in carrying on the benevolent rule ("Pre-Qin Calligraphy").

A fourth use of bronze inscriptions, according to Peng et al. (1992), was to record appointments of officials by the royal family. Such records often followed a particular format: first it was the date, month, and year of appointment as well the particular time of day; then there was the location; what followed was typically the rituals for the announcement of the appointment, including the specific details of the whole procedure as well as the content of the appointment; finally, also recorded would be the appointee's acceptance of the appointment and his gratitude toward the royal family.

As a fifth use, bronze inscriptions served as records of imperial mandates. These often were recorded alongside appointments of officials. They typically included detailed advice and admonitions given by the emperor when conferring official titles. Such records usually also included the time and place where such imperial mandates were given.

A sixth use of bronze inscriptions was to serve as legal documentation of agreements and contracts. Records unearthed include documentation of various political as well as financial disputes concerning such areas as land transactions, financial compensations, ransoms for personal freedom, and handlings of criminal cases.

A seventh use of bronze inscriptions was to sing praises of ancestors as a way of filial piety. Typical contents of such inscriptions included accounts of all the glories and honors of the ancestors, which often concluded with the bronze ware maker's prayers for ancestors' protection of their posterity.

An eighth, and last, use of bronze inscriptions was to record the laws, regulations, and decrees by the governments at various levels (Peng et al. 1992 p. 341–346).

Obviously, these bronze inscriptions were considerably more advanced linguistically, technologically, and artistically compared with the oracle inscriptions in the Shang dynasty. An illustrative feature of such advancement is that character formation in bronze inscriptions was more standardized. Characters evolved further away from the pictographs in oracle inscriptions to more abstract forms in bronze inscriptions. The number of cases of one character with multiple variations, which was pretty common in oracle inscriptions, decreased, and many rarely-used characters in oracle inscriptions disappeared in bronze inscriptions (Peng et al. 1992 p. 338–339).

A second distinguishing feature of bronze inscriptions is the increase in the number of pictophonetic characters, characters with one part indicating meaning and the other sound. Such pictophonetic formation of characters is considered more advanced than pictographic or associative-compound formations since it incorporated, for the first time, a sound-denoting component. In oracle inscriptions, pictophonetic characters comprised only a small percentage of the total characters, with pictographs and associative compounds forming the bulk (about 80%) of the entire vocabulary, whereas in bronze inscriptions pictophonetic characters were the leading category (Peng et al. 1992 p. 338–339).

From a syntactic point of view, bronze inscriptions were also more advanced in that the vocabulary became larger and more varied. First, interjectory words, not available in oracle inscriptions, began to be widely used, both at the beginning and at the end of sentences, rendering possible the expression of more complicated emotions. Second, third-person denotation, also unavailable in oracle inscriptions, appeared, enabling more accurate accounts of events. Third, bronze inscriptions saw an increase in abstract words as well as words denoting units of measure, thus making complex descriptions possible. Fourth, the length of inscriptions increased considerably. Most oracle inscriptions contained only a few characters, with the longest ones at a few dozen. In contrast, bronze inscriptions with hundreds of characters on one piece were common occurrences, allowing much more detailed narrative accounts. In addition, rhyming schemes began to be employed in many inscriptions, giving these texts a more artistic flavor (Peng et al. 1992, p. 339–340). Overall, bronze inscriptions represented a much more systematic script than oracle inscriptions, which was a natural result of the significant advancements in almost all areas of the society in that historical period. An examination of the various social, economic, and political factors that contributed to the rhetorical context for the development of both oracle and bronze inscriptions, therefore, should be revealing.

The Rhetorical Context

The use of oracle and bronze inscriptions spanned a period of more than 10 centuries (from the mid Xia Dynasty through the entire Shang Dynasty and up to at least the mid Zhou Dynasty). The sheer length of the period presents a challenge for our examination of the social context for the development of these two scripts and their roles in the evolution of the Chinese civilization. Nevertheless, some observable factors can be identified that will help us in our understanding of what prompted the use of tortoise and animal bones and bronze containers as the major writing media and technologies of the period.

Exigency—Divination and Social Systemization

To understand the exigency that prompted the use of tortoise and animal bones in the Xia and Shang dynasties and bronze ware in the Shang and Zhou dynasties, a contextualization of this historical period is necessary. Accounts of Chinese history typically begin with the primitive society, which has been traced to the Paleolithic Period. Archaeological discoveries unearthed many fossils that are believed to be cultural relics of the Beijing (Peking) Man, the Dingcun Man, the Hetao Man, and the Upper Cave Man. These cultures have left evidence of the use of simple tools and primitive social organization (Jian, 1979, p. 1–4). The subsequent Neolithic Period, which began about 5,000 BCE, witnessed several cultures including the Yangshao Culture, the Longshan Culture, and the Neolithic Culture. This period was characterized by the clan culture and represented a more organized social structure.

Oracle and Bronze Inscriptions

The next period, the slavery society, started around 2,200 BCE, when the Xia Dynasty began. The rise of Xia marked the ending of the clan social structure and the beginning of the slavery nations and the hereditary ruling system. The Xia Dynasty lasted 471 years (Jian, 1979, p. 13) until about the sixteenth century BCE, when it was terminated by Shang Tang, who established the new dynasty, Shang (Xia et. al, 1979, p. 356).

The Shang Dynasty went through 31 rulers and 496 years until about the eleventh century BCE (Jian, 1979, p. 17). The significance of the Shang Dynasty lies in the fact that its starting date of around 1600 BCE marks the first page of Chinese history that has solid archaeological backing, mostly in the oracle inscriptions (Huang, 1990, p. 6). The ensuing Zhou Dynasty (1066 BCE-221 BCE) marked the real beginning of the systemization in government, education, agriculture, commerce, and many cultural practices.

The use of tortoise and animal bones as the medium of writing in the Xia and Shang periods found its exigency in the divination practice prevalent in these two dynasties. The early part of this historical period, which coincides with the emergence and the ensuing prevalence of oracle inscriptions, witnessed the flourishing of farming, animal domestication, handcrafting, commerce, and trade. While all such activities grew out of necessity and evolved into a way of life, many factors surrounding these activities still remained a mystery to and beyond the comprehension of the Shangs. As a result, the Shangs developed a religious practice called divination. Due to the unpredictability of many things, such as the weather that affected farming, the wars with many small, neighboring nations, and the diseases and illnesses that affected people, the Shangs relied heavily on divination, which became a popular custom. This widespread practice became the exigency that dictated the need for a tool to record all the divinations. One thing that supports such a claim is that up until the Shang Dynasty, there had been no written record of any sort of the cultures that had existed. The Xia Dynasty, which immediately preceded Shang, developed a rather systematic social structure. However, there was no technology available to record its cultures. When civilization became more sophisticated and rather well developed in the Shang dynasty and the need to record the civilization became more apparent, the exigency for the invention of a writing technology thus came as no surprise.

This social trend toward systemization became a more compelling force that prompted the emergence of bronze inscriptions in the early Shang Dynasty and solidified bronze ware's status as the writing medium in the latter part of Shang and the early period of Zhou. This emergence of bronze ware as the writing medium had much to do with the improved technology in bronze making and a sharp increase in the use of bronze weapons in the Shang Dynasty. Prior to the Shang period, weapons were made chiefly of bones and stones. As the Shang Dynasty developed its military force as a result of increased warfare, weapons of greater power and better durability were in demand. The improved technology in bronze making made mass production of bronze weapons possible. The late Shang saw not only a sharp increase in quantity but also a more complete arsenal in bronze weaponry, which consisted of contact-fight weapons (such as daggers, spears, swords, and knives), distance-fight weapons (such as bows and arrows), and body armors (such as helmets and shields). For example, one archaeological dig in Anyang unearthed 730 bronze spears, 72 bronze daggers, and 141 bronze helmets, which dated back to the Shang period (Zhong 1992, p. 92). By the Zhou period, stone and bone weapons had been completely phased out (Zhong 1992, p. 93). This widespread use of bronze for military purposes naturally extended to other fields such as agriculture, commerce, art, education, and religion, and naturally led to the selection of bronze ware as the writing medium of the period.

Ideology—The Dominance of the Divine God

Though none of the systematic ideologies mentioned in Chapter 1, such as Confucianism, Taoism, Buddhism, etc., existed in the Shang and early Zhou dynasties, there was a strong religious superstition prevalent during that long period. The religious concept of the divine God was already well accepted. The Shangs believed that the God was the supreme ruler of the world, who had his own subjects, who was the dominator of nature, and who had the power to bestow either fortune or misfortune upon humans (Jian, 1979, p. 27). This God was nothing more than the fictional construction of the image of the Emperor of Shang. In addition, the Shangs also worshipped their ancestors and believed that their forefathers, especially their ex-rulers and ex-officials, could go to heaven to serve the God. In some way, therefore, their ancestors would acquire certain godly powers.

Such a religious belief dominated the thinking of the Shangs, who therefore often held grand ceremonies of sacrificial rites to their ancestors and past emperors. This worship of the God, of the natural elements the God dominated, such as the sun and the moon, the stars, the rivers, the land, etc., and of the ancestors, led to the widespread practice of divination, which became the sacred guide for people's actions. A deciding factor in the midst of all this, however, was the direct endorsement and facilitation of the concept of the divine God by the Shang rulers, whose ulterior motive was clearly to protect their reign through the powers of the God. It was no wonder, then, that divination became an essential part of the Shangs' lives.

Participants—The Ruling Class and the Populace

Records show that divination was a widespread practice among the Shangs (Jian, 1979), and there were two groups of participants in the act of divination. One was the ordinary people, the other the rulers. For the ordinary people, divination meant two things. First, it was their way of worshipping the God and their ancestors. Second, due to their lack of knowledge about the natural elements like the weather, divination was the only way for them to predict the outcome of an event. For the second group of participants in the act of divination—the ruling class, especially the emperor—divination was used in three different ways: to worship the God and their ancestors, to predict the outcomes of events, and, unknown to the populace, to underscore their rule over the people. In addition to these two groups of participants, there was an interesting third group—witches, fortune-tellers, spiritualists, and sages—people who engaged in divination as their profession. Most of these people became high-ranking government officials and were considered the possessors of knowledge, dominating the entire cultural landscape from the field of astronomy to personnel affairs (Bu & Zhang, 1992, p. 393).

Knowledge Creation—Interpreting the Divine God

The dominating rhetorical act in the emergence and development of both oracle and bronze inscriptions was obviously divination. However, divination was more than a simple act of inscription to worship the God and the ancestors or to predict the outcome of an upcoming event. Since divination was of the status of guiding and determining people's actions, it in essence was an act of creating knowledge—knowledge about the meaning of various events. Another aspect of knowledge creation was the interpretation of both the language and the writing medium. In the case of oracle inscriptions, both the language (the inscriptions created) and the use of the writing medium (the tortoise shells and animal bones) were interpreted to be, or at least associated with, the act of the God. In the case of bronze inscriptions, since both the language and the writing medium were more the result of deliberate human production and craft, both were interpreted as the means to honor ancestors and the divine God and thus to interpret the God's will. This knowledge creation act, in addition, was very pervasive since divination was used in virtually all areas of life.

Knowledge Access and Control—Powers and Privileges

Though divination was used in similar ways by both the ordinary people and the rulers in the Shang dynasty, there was a fundamental difference between these two groups of participants in their purpose for using divination. For the ordinary people, divination was a tool to guide their daily actions, such as farming. For the rulers, however, divination was not only a tool to guide their actions, such as hunting or waging a war, but also a tool to control their people. Since the social structure of the Shang dynasty incorporated a slavery system, to control the slaves and to have them serve the ruler were also the primary focus of the emperor. Thus the emperor could, for example, interpret the word *zhong* (people) as referring to the slaves and *wang* (emperor) as himself. In this sense, divination (and the inscription) was more a control tool than a practical means of communication.

With oracle inscriptions, access to the writing medium seemed non-exclusive as tortoise shells and animal bones were obviously readily available to the ordinary people, but the use of the medium was limited mostly to the ruling class. In the invention and use of bronze inscriptions, access to the written communication medium was extremely exclusive and almost totally limited to the rulers and the aristocrats. This exclusive access resulted in the use of bronze inscriptions as the means to communicate with the God and ultimately to interpret the God's will to legitimize the rulers' reign.

The Communication Medium—Language as Both the Object of Invention and the Medium for Communication

The communication medium in the case of oracle inscriptions was rather special. In other cases of technology development, the communication medium basically refers to the language and how that language is used to communicate the meaning created. In the case of oracle inscriptions, the language itself was the object of invention (i.e., the language *was* the technology). In addition, the act of using the material medium to carry that language (i.e., the tortoise shells and animal bones) was an invention as well. Thus, language became both the object and the medium of the technological invention.

However, such a recognition of the role of written language and media in both oracle and bronze inscriptions should not blind us to another medium that people used to communicate knowledge—the oral language. Though in the Shang period the written form of language was in its primitive stage of invention, the oral form of Chinese had already been in use for centuries. In this case, it acted as the primary medium of communication.

However the oracle and bronze inscriptions were used and, despite the fact that both were only secondary forms of communication to the oral medium, they represented a first major breakthrough in the development of writing and writing technologies in China. In many ways, this important milestone spearheaded a series of writing technology inventions that were to come in the next couple of millennia.

5 Early Forms of Pen, Ink, and Paper

The pre-Qin Chinese culture was ... neither a singular aristocratic culture nor a closed body of unchanging traditions throughout Xia, Shang, and Zhou dynasties. Rather, it was a pluralistic amalgamation of traditions integrating the cream of multiple ethnic cultures, rising above the limitations of local and clan cultures while sustaining their distinguishing characteristics. The pre-Qin society was like a big melting pot, meshing together a plurality of cultures, ideologies, customs, and traditions.... (Peng, 1992, p. 1)

Chinese writing as a special form of art has been developed through the use and improvement of various tools and implements since ancient times. Paper, brush, ink, and ink-slab, the "four treasures of a scholar's study," have been the basic implements for committing thoughts to writing. (Tsien, 2004, p. 175)

There has been much discussion but only fragmentary pieces of evidence about the early forms of pen, paper, and ink used in China. Most of these early forms of writing technologies either started or were further developed in the Zhou dynasty, a historical period immediately succeeding the Shang dynasty. Starting in the eleventh century BCE and ending in the third century BCE, the Zhou dynasty lasted about 900 years (Lattimore, 1946, p. 31). It is divided into two periods: the Western Zhou (the eleventh century BCE to the eighth century BCE) and the Eastern Zhou (the eighth century BCE to the third century BCE). The Eastern Zhou is further divided into two sub-periods:

Early Forms of Pen, Ink, and Paper

Spring and Autumn (770 BCE to 476 BCE) and the Warring States (476 BCE to the third century BCE). The Zhou dynasty was a historical period of war, change, and rapid developments. During the first part of this dynasty, bronze inscriptions were the main writing form as they were naturally extended into this period from the preceding Shang Dynasty. At the same time, the Zhou Dynasty also witnessed the emergence of several important forms of writing technologies, namely, bamboo pens, soot ink, bamboo and wood slips as primitive forms of "paper."

These writing technologies were then significantly advanced during the ensuing Qin Dynasty (221 BCE to 206 BCE), which was a very short one, especially when compared with the preceding Zhou Dynasty. However, we cannot afford to ignore this historical period for the main reason that Qin was a dynamic period of social development, which shaped the development paths of these writing technologies.

PEN

A discussion of writing naturally invites questions about the type of pens used and their origin. Unfortunately, due to the lack of direct archaeological evidence, this is, for the most part, still a mystery. For example, although the invention of the brush pen is generally attributed to Meng Tien, a general in the Qin dynasty, in the year 250 BCE (e.g., Hunter, 1947, p. 4), there is evidence that strongly suggests "that the brush may already have had a long history before the time of Meng Tien" (Carter, 1955, p. 8–9). Carter claims that "archaeological proof is given by the evidence of brush writing on Yang-shao pottery (*ca.* 2000 BCE), on jade, and on oracle bones of the late Shang period (*ca.* 1300–1028 BCE)" (p. 9). This considerably pushes back the invention date for the brush pen. Other researchers seem to agree with Carter, more or less. William Boltz (1999), for example, claims that "evidence of brush writing in the late Shang period ... is known from characters that seem to have been written with a brush on the smooth surface of pottery or jade ..." (p. 108). Robert Bagley (1999) also claims that there is clear evidence "for brush writing and, less direct, for wood or bamboo writing surfaces" in the Shang dynasty (p. 182).

Some (e.g., see *General Knowledge,* 1980) have even suggested that some of the pottery unearthed from the Neolithic age carried certain patterns that are strongly suggestive of writing brush strokes (p. 280). In addition, many instances of un-inscribed characters have been found on animal bones from the Shang period, and the smooth strokes of these characters seem to suggest they were probably written with brush pens. Rawson's (1999) research of the Zhou Dynasty also supports such a claim:

> ... on the evidence of early finds of inscribed bronzes, the Luoyang area was inhabited by eminent members of the Zhou elite. Specialized scribes and artisans must presumably have been involved in writing and casting inscriptions that describe their activities. We know that scribes were using brush and ink because of some weapons and a vessel that have ink inscriptions upon them [. . .]. (p. 405)

However, all these guesses at the origin date for the brush pen, however educated, lack archeological backing. The earliest actual brush pens discovered so far date back to the Warring States period (476 BCE to the third century BCE). One of these brush pens, unearthed in Changsha, Hunan Province, is made of bamboo and has a thin shaft of 0.4 centimeters in diameter and 18.5 centimeters in length, with the brush part made of 2.5-centimeter-long, high-quality rabbit hair (*General Knowledge,* p. 280). Tsien's (2004) research also documented several discoveries of brush pens from as early as Warring States to as late as Qin (221 BCE to 206 BCE) and Han (206 BCE to 220 CE), with most of these pens made of either wood or bamboo for the shaft and different kinds of animal hair for the brush (p. 178–182).

Invented probably in the Shang period and further developed and popularized in the Zhou dynasty, the writing brush was a significant invention in the development of writing technologies. This invention, according to Carter (1955), "worked a transformation in writing materials, [which was] indicated by two changes in the language," one being that "the word for chapter used after this time means 'roll'" and the other being that "the word for writing materials becomes 'bamboo and silk' instead of 'bamboo and wood'" (p. 4). Such a role was also acknowledge by Hunter (1947) who thinks that this innovation "not only revolutionized the writing of Chinese characters, but was instrumental in the further development of woven cloth as a writing material, a substance which [sic], along with the papyrus of Egypt and the parchment of Asia Minor, made possible the manuscript scroll, the first form of book in its true sense" (p. 4).

Ink and Ink Slabs

The origin of ink was as much a mystery as was the invention of pens. Though the invention of ink is traditionally credited to a famous calligrapher and ink maker named Wei Tan in the third century CE, there is evidence that ink was invented much earlier. As Tsuen-hsuin Tsien (2004) points out, "the arbitrary dating of the use of ink around the third century CE is apparently speculation and contrary to the testimony of both early literature and later archeological discoveries" (p. 182). Carter (1955) believes that ink "may have been known in classical times, possibly even in the Shang period" because, for example, "the writing on the *Chu shu chi nien,* dating from 299 BCE, was in ink" (p. 32). Tsien's following account is quite revealing in establishing the use of ink well before the third century:

> Mencius (372–289 BCE) speaks of a carpenter's "string and ink"; ink was probably used for writing in his time. The *Chuang tzu* mentions that when Prince Yuan of Sung desired to have a picture painted, all the scribes of the court stood up "licking their writing brushes and mixing their ink." Chou She, the counselor of minister Chao Yang (d. 458 BCE) of the Chin state, said to his master: "I wish I could be your critical subordinate, handling tablets with brush and ink and watching after you to record whatever faults you may have." The *Kuan tzu,* a collected work attributed to Juan Chung (d. 645 BCE) but probably written at a later date, mentions that Duke Huan of Ch'I (r. 685–643 BCE), in regard to the improvement of his administration, "asked the officials to record his orders on a wooden board with ink and brush." (p. 182–183)

One thing is clear: whenever brush pens are used, there has to be ink to be used with them. The evidence of un-inscribed written characters from the Shang Dynasty suggests ink was used at least as early. What is not clear is whether ink used to write those un-inscribed characters found on pottery and animal bones from the Shang period was directly from the natural minerals or a product of synthesis from soot. An archaeological digging of a tomb from the Warring States period in Changsha, Hunan Province, in 1954 unearthed some bamboo slips that carried words written in ink; alongside these bamboo slips was also found a bamboo basket filled with a dark-colored soil-like substance, which most likely was the ink used to write on those bamboo slips (*General Knowledge,* 1980, p. 283). However, analyses of the ink used on silk painting and books from the Warring States period indicate more advanced ink-making technology than what the dark-colored soil-like substance would suggest.

One popular form of ink was lampblack, a fine soot that "was obtained by burning certain kinds of woods or liquid, including pine, tung oil, petroleum, and probably lacquer (Tsien, 2004, p. 184). Tung oil and petroleum were used in ink making much later than the Zhou period, with tung oil being used in the tenth century and petroleum in the eleventh century. For the Zhous, "Pinewood was probably the most popular material for obtaining lampblack, and pine soot is still the best material for making black or 'India' ink today" (Tsien, 2004, p. 184). One old recipe for mixing ink included in the *Qimin Yaoshu,* a work on agriculture and manufacturing written by Jia Sixie of the fifth century CE describes the ink-making method as follows:

> Fine and pure soot is to be pounded and strained in a jar through a sieve of thin silk. This process is to free the soot of any adhering vegetable substance so that it becomes like fine sand or dust. It is very light in weight, and great care should be taken to prevent it from being scattered around by not exposing it to the air after straining. To make one catty of ink, five ounces of the best glue must be dissolved in the juice of the bark of the *cen* tree which is called *fanji* wood in the southern part of the Yangtze valley. The juice of this bark is green in color; it dissolves the glue and improves the color of the ink.
>
> Add five egg whites, one ounce of cinnabar, and the same amount of musk, after they have been separately treated and well strained. All these ingredients are mixed in an iron mortar; a paste, preferably dry rather than damp, is obtained after pounding thirty thousand times, or pounding more for a better quality.
>
> The best time for mixing ink is before the second and after the ninth month in a year. It will decay and pro-

> duce a bad odor if the weather is too warm, or will be hard to dry and melt if too cold, which causes breakage when exposed to air. The weight of each piece of ink cake should not exceed two or three ounces. The secret of an ink is as described; to keep the pieces small rather than large. (cited in Tsien, 2004, p. 185)

The resulting form of ink, as indicated in this description, is solid and is often of rectangular shape. How, then, is solid ink used for writing? The answer lies in another piece of writing technology that always goes hand in hand with this kind of ink—the ink slab.

The ink slab is a palette, often made of polished stone, used as a platform for rubbing the ink stick. The ink slab typically has a flat surface with the edges slightly elevated to hold water on the surface. An ink stick is then rubbed against the surface of the ink slab, during which process ink from the ink stick dissolves and mixes with the water until it forms a thick, almost sticky fluid. A brush pen is then used to dip gently into the thick ink. With the hair portion of the pen holding a small amount of ink, the brush pen can then be used to write on various surfaces.

Archaeological excavations from the Neolithic period have unearthed pottery with various color patterns drawn with pigments of very fine quality, which suggests some kind of ink-rubbing tool might have been already available during that period. More substantial archaeological evidence for the existence of ink slabs has been identified based on the discovery of brush pens, ink, and ink drawings on silk and bamboo unearthed from several tombs from the Warring States period in Hunan, Hubei, and Henan, which convinced archaeologists of the use of ink slabs in the Zhou Dynasty. However, the first concrete evidence was a stone ink slab found in a Qin tomb in Hubei, which was obviously the result of rough crafting from natural pebble and carried visible evidence of ink rubbing as well as ink marks (*Common Facts*, 1980, p. 286).

Early Forms of Paper: Bamboo, Wood, and Silk

The use of pen, ink, and ink slabs naturally entailed the use of some form of paper materials to write on. Throughout history, the Chinese explored various materials for writing. As Hunter (1947) attests,

> Long after the period of drawing in the sands and upon the walls of caves, came the more workable materials such as wood, metal, stone, ceramics, leaves, barks, cloth, papyrus, and parchment as basic surfaces upon which to incise or inscribe hieroglyphics and characters, each of these substances, for want of a more flexible and pliable material, faithfully fulfilling its individual requirements through the centuries. (p. 3–4)

Traditional records typically attribute the invention of paper to Cai Lun in CE 105. This date is not inaccurate if paper is meant in the modern sense. However, evidence abounds that earlier forms of "paper," such as bamboo, wood, and silk, were already in use long before the second century. This definition of paper is certainly debatable. By including bamboo, wood, and silk in the category of paper, we are not talking about paper made *from* bamboo, wood, or silk, but *of* these materials. Such a broad definition of paper is not too far fetched as it is accepted by many researchers.

According to Tsien (1962), the history of paper use in China can be divided into three periods: (1) bamboo and wood from the earliest times to the third or fourth century CE; (2) silk from the fifth or fourth century BCE to the fifth or sixth century CE; and (3) paper from the second century CE to the present time (p. 91). Such period designation locates the use of bamboo, wood, and silk as early forms of "paper" as early as at least the Zhou Dynasty. For example, in describing a bronze inscription for a man named Song in about 825 BCE, Shaughnessy (1999b) notes that "this inscription shows that prior to the audience, the court scribes had prepared Song's appointment in writing—written on bamboo or wooden strips" (p. 298–299). We have reason to believe that writing was becoming an important part of the social life in the Zhou period. According to Carter (1955), in the classical period through the Zhou dynasty, ancient Chinese used the wood for short messages and bamboo for longer ones and for books (p. 3–4). Here is an account by Carter of how wood and bamboo were used:

> The bamboo was cut into strips about nine inches long and wide enough for a single column of characters. The wood was sometimes in the same form, sometimes wider. The bamboo strips, being stronger, could be perforated at one end and strung together, ei-

ther with silken cords or with leather thongs, to form books. (p. 4)

The earliest forms of books in China were *jian ce* (简策) and *ban du* (版牍), both meaning books or booklets. The former, *jian ce* (简策), refers to bamboo or wooden strips strung together. A single strip of bamboo or wood is called *jian* (简), which usually holds only one line of words. When many such strips are strung together, they form *ce* (策). One variation of this word is *ce* (册) (pronounced the same), which clearly resembles bamboo or wooden strips strung together and is still used today to mean books. *Ban du* (版牍), on the other hand, refers to thin wooden blocks only. A wooden block with no words on it is called *ban* (版), while one with words is called *du* (牍). *Ban du* (版牍) is used to hold a short article, often fewer than a hundred words while longer articles would require the use of 简策. Writing on these bamboo and wooden strips and blocks was done with brush and ink. When characters were written wrong, they would be scraped off with a knife (*Common Facts,* 1980, p. 288–289).

Common Facts claims that bamboo and wooden booklets existed as early as the Shang Dynasty and became the main form of books from the Warring States period, through the Qin Dynasty (221 BCE to 206 BCE), to the Han period (206 BCE to 220 CE) (p. 289). Numerous wooden and bamboo books have been unearthed in several archaeological digs. For example, one dig of a Qin tomb in Hubei in 1975 unearthed more than 1,000 bamboo strips, which were mainly recordings of Qin's laws and regulations. Although the thin strings used to thread the bamboo strips together were already decomposed, it could be observed that the bamboo strips were strung together on the top, in the middle, and at the bottom. A more important archaeological discovery was made in 1972 near Linyi, Shangdong, where more than 4,900 bamboo strips were found, among which were some well known works by various Qin scholars such as *Sun Wu's Art of War* (孙子兵法), *Sun Bin's Art of War* (孙膑兵法), *Yu Liaozi* (尉缭子), *Six War Strategies* (六韬), *Guan Zi* (管子), and *Yan Zi* (晏子). Of these, *Sun Bin's Art of War* had been missing for more than 1,700 years, and this discovery was of historic significance in archaeological research about the Qin Dynasty (*Common Facts,* 1980, p. 289–290).

According to Peng et al. (1992), based on the bamboo booklets unearthed from the Warring States period, the content on these bamboo booklets fall into four categories: (1) ancient books, (2) records of memorial ceremonies, (3) court documents, and (4) inventory records, while Qin bamboo booklets were mainly recordings of legal documents, plus some chronicles, daily events, and sex techniques (p. 359). Wooden tablet excavations from the Zhou and Qin periods have been rare; only a total of four have been discovered so far, two being family letters and the other two the recordings of a Qin legal document called *Land Regulations* (p. 359).

Bamboo and wooden booklets were heavy and cumbersome and were hard to flip through or carry. Qin Shihuang, the emperor of Qin, after unifying China, had to read more than a hundred pounds of such booklets every day. Legend has it that a Western Han literary scholar Dongfang Su wrote a memorial to the Wudi Emperor that used three thousand bamboo strips; the booklet was so heavy that it had to be carried into the palace by two people (*Common Facts,* 1980, p. 291)

Bamboo and wood hold a special place in the history of writing technology development. First, though writing had existed on shells, bones, stones, and bronze long before bamboo and wood were used, it was not until the use of bamboo and wood strips that the concept of book was actualized. As Tsien (1962) has argued, "the direct ancestry of the Chinese book is believed to have been the tablets made of bamboo or wood which [sic] were connected by a string and used like a paged book of modern times" (p. 90). A second notable aspect about bamboo and wood is that they "were not only the most popular materials for writing before paper was invented but also served probably a longer period in Chinese history than any other material as a medium of writing" (Tsien, p. 90). The fact that they survived for about three centuries after paper was invented was proof in itself of how important they were in the history of Chinese civilization.

Another major early form of paper that can be traced to the Zhou Dynasty is silk. The use of silk as a material for clothing, musical strings, exchange medium, etc. has long been recorded. The exact starting date for silk to be used as a writing material is not clear. However, Tsien (1962) believes that "the use of silk for writing originated no later than the sixth or seventh century BCE and continued even after the third or fourth century CE" (p. 116). His claim is backed up by both ancient literature and recent archaeological discoveries (p. 117). Hunter (1947) also believes that silk's use as a writing material dated back at least to the fifth to fourth century BCE, possibly even earlier (p. 465). Peng et al. (1992) believe silk as a writing material was already commonly used by the Warring States period. One example is the well-known silk book from Chu, one of the states in the Warring States period, unearthed in 1942 and now kept in the United States. It contains about 900 words, divided into fourteen paragraphs, with two of them placed in the middle and the other 12 around the edges. Its content involves astronomy, myths about the four seasons, and various taboos (p. 360).

Obviously, silk as a writing material was developed as an alternative to bamboo and wood. Compared with the bulkiness and heaviness of bamboo and wood, the light weight of silk made it an attractive alternative. In addition, silk was soft, durable, and absorbent, which made it easier to preserve or carry. However, one thing prevented silk from completely replacing bamboo and wood as the main writing material: its cost. Due to the scarcity of the raw material and the sophistication involved in processing silk to make it paper, it was much more expensive to make paper out of silk than of bamboo or wood. Therefore, silk had always remained a luxury alternative to bamboo and wood papers and "was used only when bamboo and wood did not suit the special purposes" (Tsien, 1962, p. 127).

Despite its relatively high cost, silk was still quite well used. Tsien (1962) identifies several different uses of silk in writing. While bamboo tablets were often used for preliminary drafts because it was easier to make changes on bamboo, silk was more often used for final editions of books. A second use for silk was "for illustrations appended to books of bamboo tablets" (p. 128). For example, a book by Sun Wu of the sixth century BCE contains eight rolls of illustrations, and another by Sun Bin of the fourth century BCE has four rolls (p. 128). Obviously, silk was a better material for drawing than the narrow strips of bamboo. A third use for silk paper was for map drawing. Just as with illustrations, it was much easier to draw maps on silk than on bamboo. A fourth use for silk was "for inscriptions for sacrifice to spirits and worship of ancestors" (p. 129). A fifth use was for recording sayings by kings and royal houses so that they could be passed on to later generations. Another use was "for permanent records of exceptional honors awarded to great statesmen and brilliant heroes of military achievements in the government" (p. 129). As is shown by Tsien, most of these uses were in practice in the late Zhou dynasty, namely, the Warring States period, and were extended to the Qin Dynasty.

Writing technologies in the late Zhou and the ensuing Qin periods were not limited to these media. In addition to writing on bamboo, wood, and silk, several other forms of writing were identified with the Warring States period, and the ensuing Qin Dynasty as well, and these employed different media:

- Bronze inscriptions—Inscribed on bronze vessels, these were obviously inherited from the Shang and earlier Zhou, as discussed in the previous chapter.
- Weaponry inscriptions—Words were also found on various weapons. These inscriptions mainly recorded the names of the weapon makers and the officers in charge.
- Coin inscriptions—These resulted from the emergence and use of metal coins. Due to the small surface of these metal coins, words on these coins were often very simple, usually denoting the place of make, their weight, and their value.
- Royal decrees—When the emperor needed to send out a decree (e.g., to declare war) it would be carried by people to send it to the right person. These decrees were written on various media, including gold, bronze, jade, animal horns, bamboo, wood, lead, etc. Often in the shape of tigers, they were typically made into two separate pieces, to be held by two separate people. When needed to exert their power, the two pieces would be brought together; only when they perfectly match and complement each other in shape would it be considered authentic and authoritative.
- Units of measurement—Toward the end of the Warring States period and the early Qin Dynasty, when Qin Shihuang was unifying China, measuring tools were developed that employed uniform measurement units to avoid confusion.
- Jade and stone inscriptions—These often recorded oaths of alliance or loyalty.
- Imperial seal inscriptions—Imperial seals in the Warring States period carried people's names, official titles, or blessings. These seals were made of various materials such as bronze, jade, agate, and stone.
- Pottery inscriptions and writings—Pottery making techniques were rather advanced in the Warring States period. As pottery making became more specialized and commercialized, pottery

inscriptions and writings because more common. These inscriptions and writings mainly recorded the names of official titles, places, the pottery, the person who made it, and the person who would use it.
- Stone inscriptions—Archaeological digs have rarely unearthed stone inscriptions. So far, only three pieces have been identified, but the characters on one of these pieces appear exquisite and finely written and are of great value to the research into the language of the Warring States period.(Peng et al., 1992, p. 359-362)

The co-existence of so many different writing forms and media in late Zhou and the subsequent Qin periods was unprecedented. It greatly spurred the development of the Chinese language. While the total vocabulary was no more than 1,000 for oracle inscriptions and fewer than 2,000 for bronze inscriptions, it reached 3,300 in bamboo, wood, and silk booklets (Peng et al., 1992, p. 363). The development in language was at once a result of and basis for cultural and social advancement.

The Rhetorical Context

That so many different writing technologies were invented or expanded into full fledged use in the Zhou and Qin Dynasties should not come as a surprise if we consider the fact that the Zhou Dynasty spanned about 900 years. More importantly, this dynasty, especially the Spring and Autumn period and the Warring States period, and the ensuing Qin Dynasty witnessed spurs of advancement in political, economic, and especially cultural areas.

Exigency—Burgeoning Social Progress

If the Shang Dynasty was characterized by the slavery system, then the early part of the Zhou Dynasty is traditionally considered to be the beginning of the feudal system, which, by the end of the Zhou Dynasty in the third century BCE, had evolved into a well developed, highly sophisticated system. Not only were there major political reforms, such as the Shangyang reform in the fourth century BCE, but there were significant advancements in education, military science, literature and art, religion, agriculture, and commerce. The late Zhou period, including the Spring and Autumn and the Warring States periods, saw the decline of the imperial regime and a rising contention for supremacy by seven big states of power. Amid all this contention and the pursuit of supremacy, the big powers saw a pressing need for political reform as a necessary step to strengthen themselves. A natural result of these political reforms was the promulgation of numerous new laws and regulations by these states.

These important developments in almost all areas of the society inevitably created an exigency for writing and new writing technologies. The dissemination of new laws and regulations dictated the need for more effective communication forms and media, and the development of existing writing media and the invention of new media thus came as no surprise. As Loewe (1999) points out, "the growing sophistication of government during the Warring States period demanded convenient means and materials for compiling the increasingly large number of documents that officials were being required to handle" (p. 1010). Under such a demand, the enhancement in pen making and paper making resulted in better brush pens and improved paper, such as the silk paper.

Ideology—"The Contention of a Hundred Schools of Thought"

A more revealing change lies perhaps in the rapid development and the multiplicity of the cultural ideologies of the time. From the Shang and Xia up until the Spring and Autumn period, divinity and the God's will had been the dominating ideology, which, to a large extent, dictated how oracle and bronze inscriptions were used. By the Spring and Autumn period, the concept of divinity had waned in its power and began to be challenged. Debates about the God vs. humans led to a new cultural ideology that separated the God from humans and placed priority on the latter over the former (Bu & Zhang, 1992, p. 403) and provided a breeding ground for the multiplicity of ideologies of this period.

The most notable of these ideologies was no doubt Confucianism, which emerged as arguably the most influential philosophy in the history of Chinese thought. The central tenets of Confucianism included a humanity-based philosophy, ethics and moral values grounded in the concept of benevolence, and the doctrine of the mean that emphasized balance between opposing forces. Of particular interest to our discussion of writing technologies here is the Confucianist epistemology, which emphasized both learning and reflection. With the unrivaled influence of Confucianism on the Chinese culture of the period and subsequent periods, this emphasis on learning and reflection, in turn, prompted the start of the tradition of learning and a plethora of educational activities that inevitably drove the invention of various writing forms and technologies.

The ensuing Warring States period witnessed even greater intellectual prosperity, resulting in multiple schools of thought, each represented by one or two well-known philosophers. There was Mo Zi, who proposed the ideas of valuing talent in choosing officials, equality between the aristocrats and the populace, love for all, peace, etc. Then there was Lao Tzu and Zhuang Zi, who developed Taoism, which advanced the first systematic philosophy on the origin and evolution of various phenomena of nature. There was Mencius, who furthered and perfected Confucianism. There was Xun Zi, who integrated principles from several schools of thought, including the Taoist perspective on nature and the Confucianist idea of etiquette. There was Han Fei, who established Legalism, which advocated a balanced and fair legal system. Then there was also Sun Zi, who wrote the famous work on military stratagem, *Sun Zi*. These philosophers and the works they created exerted far-reaching influence on the politics and culture of the feudal society and no doubt shaped people's ways of thinking and their perspectives on many things (Jian, 1979, p. 85–90). With the multiplicity of cultural ideologies naturally came the need to document such ideologies and to advance learning and great thought, providing a fertile ground for experimentation with writing forms and writing technologies.

Participants—The Multiplicity of Participant Groups

With the freedom in thought came more active participation by various groups of people in the development of writing technologies. Unlike the Shang Dynasty, when there were basically only two classes, the ruler and the slaves, the Zhou Dynasty, with its privatization of land and the consequent abolishment of the slavery system, created several distinct classes: the emperors, the government officials, the landlords, the merchants, the intellectuals, the peasants, etc.

Each of these groups of people had its own interests in advancing writing technologies, and for each group, the development of a particular writing technology often signified a different act with a different purpose. For the emperors, the writing technologies not only provided a means for them to pass on their records to posterity, but also was, in a way, a tool to consolidate their rule. For government officials, the development of writing materials was a way for them to secure a better status with the emperor. They acted often as the political elite, and sometimes even as the technical elite as they were put in charge of developing a certain writing material, and were thus often in a position to shape the course of development. The intellectuals mostly acted as the technical elite and were often a major positive force behind the development of writing materials, as such technological development directly benefited their literary interests. Even for the landlords and merchants, the availability of writing technologies served practical purposes as, for example, a title deed for the land provided a written testimony to the landlord's ownership of property.

Amid this multiplicity of participant groups, one distinct group was left out: the poor peasants and handicraftsmen. For these people, life was still a hard struggle and writing or education a luxury. It is hard to imagine how this group of people could define, shape, or even influence writing technology development in any form.

Knowledge Creation—Advancing All Walks of Life

Knowledge creation in the form of writing and writing technology development of this period formed a sharp contrast with that in the Shang and Xia Dynasties, when oracle and bronze inscriptions were the only dominating writing forms. Although still a privilege and luxury for many people, writing in the late Zhou period had shifted from a singular act of divination to a multi-faceted act of social progress and advancement. Writing technology development, likewise, evolved from an act of official execution controlled more by the royal administration to an act of public participation.

Apart from the influence of burgeoning ideologies, much of this shift also had much to do with a new trend in the field of education—the decline of government-sponsored elite schools and the popularization of private schools. This de-elitization of education came as a result of the shift of economic and political power from the royal family to the populace, which in turn was a natural result of the advancement in the means of production that led to increased possessions by the public (Qiao & Liu, 1992, p. 378). In this de-elitization process of education emerged a special social class made up of scholars, a somewhat middle social stratum between senior officials and the common people. These scholars, who used to serve in the government, now went to various states and became teachers in private schools. Such a phenomenon contributed much to the cultural diversity of the period and ended government-controlled education.

This privatization of education put the public in a more powerful place in defining the meanings of the various writing technologies of the period and determining their development paths. These writing technologies—namely, pen, ink, ink slab, bamboo and wooden booklets, and silk—were no longer tools exclusive to the ruling class; rather, they were considered means for the common people on their path to education and knowledge.

Knowledge Access and Control—Open Access for the General Public

Unlike oracle and bronze inscriptions, the access of which was mainly limited to the royal family and some aristocrats, the writing technologies of the late Zhou and early Qin period were readily available to members of different social strata. This more open access had much to do with the fact that education in this historical period was open to everybody, at least in principle.

The privatization of educational entities allowed teachers and masters of learning to open their schools to anybody who could afford it. Confucius, for example, was one such teacher of private schooling, albeit the most famous one. In his entire teaching career, he advanced and advocated the notion of "education for everybody regardless of his social status," and his students included not only aristocrats and wealthy merchants but also common people, people of low social status, even people of violence, thieves, and criminals (Qiao & Liu, 1992, p. 378). He believed that everybody was teachable as long as he was committed to learning. His students came not only from diverse social backgrounds but also from different geographical regions including the "less civilized" minority areas outside the main states. As a result, his private school reached a considerable scope, with students numbering as many as 3,000, of whom 72 protégés became masters of several subjects.

In the meantime, probably under the influence of Confucius's approach to private education, many other famous scholars followed suit. Numerous private schools emerged, and so did many masters representing diverse schools of thought. For the first time in ancient Chinese history, education matured into a complete, systematic enterprise. This private educational enterprise, according to Qiao and Liu (1992), had some distinct characteristics. First, teaching pedagogies were unique and diverse. Second, these schools became the breeding grounds for different schools of thought. Third, they provided a forum for discussions about and participation in politics. Fourth, they were important resources for training intellectuals and other talents. Fifth, the masters would go all over China to disseminate knowledge (p.380). This open access to education and wide dissemination of knowledge rendered writing technologies accessible products to the general populace. For the first time, writing technologies became tools for the common people rather than exclusive symbols of power for the privileged few.

Communication Medium—Oral and Written Dissemination

One main form of the dissemination of knowledge in this historical period, including that of writing tools, was oral communication through teaching by those masters of private schools. As mentioned in the last section, many of these masters of private education would travel to various and distant places on their education tours. Confucius, for example, traveled many times with his followers to different states to spread his learning and thoughts. Mencius's travel entourage often included "several hundred followers and dozens of carriages" (Qiao & Liu, 1992, p. 381). Xunzi, a resident of Zhao, frequently traveled with his followers between the states of Qi, Chu, and Zhao. Such education tours had a significant impact on the dissemination of learning and knowledge. In addition, these education tours, plus the daily teaching at the private schools, taught people not only knowledge about different subjects but also inevitably the use of writing technologies such as pen, ink, bamboo tablets, and silk, which were already in use in private school education.

A second form of communication was the publication and dissemination of a large number of books, encompassing a diverse array of subject areas. In the field of literature, for example, there was *The Book of Songs*, the first poetry collection in the history of China. *The Book of Songs* collected 305 poems representing different geographical regions and spanning a period of about 500 years from the beginning of Western Zhou through the middle of Spring and Autumn. The publication of this poetry collection was in large part due to the fact that the Zhou Dynasty had court-appointed official poem collectors. These collectors would go out to various places, ringing a bell in their hands,, and collect poems from the people. They would then report these poems to the music master, who would then present them to the emperor, who used these poems as a tool to find out about what his people were thinking (Shang, 1992, p. 481). In the official, elite schools in the Zhou Dynasty, poems were used as teaching materials. This practice later spread to private schools as well and became a well-established tradition until the Qin Dynasty, when Emperor Qin initiated his infamous movement to persecute scholars and burn books.

There was also the Chu poetry, a distinctive genre of poetry developed by the Chu poet Qu Yuan, whose *Li Sao* is a masterpiece of romantic poetry. *Li Sao* consists of more than 370 lines and over 2,400 words and is divided into two parts, the first part relating the author's life, dreams, and the process of his involvement and failure in political reform, and the second part his relentless pursuit of his political dreams (Shang, 1992, p. 486). There was also the flourishing of historical prose works, including *The History by Zuo*, a comprehensive narrative-style history covering the period from 722 BCE to 464 BCE; *Guo Yu*, a collection of the famous sayings and adages by the aristocrats; and *The Rhetoric of the Warring States*, a written record of the persuasive speeches made by the advising officials and counselors of the court.

In addition to poetry and historical prose, the late Zhou Dynasty also witnessed the publication of the masterpieces representing the various schools of philosophical thought. There was *The Analects*, the masterpiece of Confucianism; *Lao Tzu*, the authoritative work of Taoism; *Mo Zi*, the representation of Mohism; *The Book of Mencius*, an incisive and eloquent work furthering Confucianist thought; *Zhuang Zi*, the collection of the thoughts of its namesake philosopher that extended Lao Tzu's Taoism; *Xun Zi*, a critique of the various schools before Xun Zi, a work considered symbolic of the maturation of pre-Qin expository prose; and *Han Fei Zi*, a work known for its strong logical reasoning (Shang, 1992, p. 495–503).

These masterpieces exerted considerable influence on the dissemination of knowledge about the various schools of philosophical thought. Although there is no concrete archaeological evidence that attests to any systematic efforts to disseminate knowledge about writing technologies during that period, evidently, the mere presence of and public access to these writing media and their daily use in private education provided the best medium of communication about the nature and uses of such writing technologies.

6 The Modern Form of Paper

> The production of books and documents in China was outstanding both in quantity and quality. Chinese classical literature is believed to rank among the greatest of all the nations. Moreover, China has the richest and most detailed historical record. From 722 BCE of the Spring and Autumn period until now, the Chinese have kept their annals almost every year. The number of words in the Thirteen Classics is several times that of the Old Testament, which is of the same genre and period. At the close of the fifteenth century CE, China probably had produced more books in terms of title and volume than all other nations put together. Even during very early times books were numerous and varied in subject matter, as we can see from the titles in ancient bibliographies. (Tsien, 2004, p. 2–3)

The publication of the large number of volumes of books in early Chinese history, as mentioned by Tsien here, rested on one important factor: the invention of paper in the modern form. As early forms of paper, such as bamboo and wooden tablets and silk, were either too cumbersome or expensive for the large-scale production of books, the invention of the modern form of paper was nothing short of a milestone, not only in the development of writing technologies but in the advancement of Chinese thought and culture. As Tsien (2004) has pointed out, the "unbroken tradition of Chinese civilization is largely due to the uninterrupted use of the ancient literature in fundamental education and also to the assiduous study of the books preserved from antiquity that was for many centuries the main way to achieve prominence in society" (p. 2). In this sense, how this important invention (i.e., that of modern paper) occurred and how it developed are worthwhile topics for exploration, especially with regard to our understanding of the continuation of Chinese traditions.

THE ORIGIN OF PAPER

Although primitive forms of paper existed as early as the Shang period, the invention of paper in the modern form, some believe, did not come about until around early second century CE As mentioned earlier, traditionally this invention is credited to Cai Lun in CE 105. This is in large part due to an account by Fan Ye in his *Hou Han Shu* (*The History of the Late Han Dynasty*) written in the Northern and Southern Dynasty (420 CE to 589 CE). However, many researchers believe that the invention of paper was more a gradual and evolutionary process that involved many people and many experiments over an extended period of time.

Several archaeological discoveries in the twentieth century point to dates prior to, and some much earlier than, CE 105 when paper in the modern form was already in existence. One archaeological dig in Xinjiang in 1933 unearthed a piece of white hemp paper from the Han Dynasty, which was of rectangular shape, about 10 centimeters long and 4 centimeters wide, and of rough quality. A later digging, in 1957, in Xian found several pieces of paper, most of which were already torn, with the biggest piece about 10 centimeters in both length and width and 0.14 millimeters in thickness. Also unearthed along with these papers were a bronze mirror and a bronze sword, and archaeologists determined the date of these paper to be no later than 118 BCE A third discovery was made in Gansu in 1973–1974, where two pieces of hemp paper from the early Han period (first century BCE) were found. One of these was 21 centimeters long and 19 centimeters wide, of clean and white color, thin and of fine quality, with one side rather smooth and the other side rather rough. Another discovery was made in Shaanxi in 1978, where some paper crumpled into balls was found to have been used as space fillers in the basin-like bronze ware unearthed in that archaeological discovery. This paper, beige in color, was made of hemp, and the fiber of the paper was rather rough. The date of this paper was determined to be prior to CE 1–5. In 1979, eight pieces of paper were found in Dunhuang, all of which were crumpled when unearthed. These papers were of varying quality and ranged in their origination date from around 65 BCE to around CE 10. A more recent discovery was made in 1986, when a map made of paper was found in Gansu.

The map paper was thin and flexible, had a very smooth surface, and had mountains, rivers, and roads drawn in thin, black lines. The drawing style was similar to that found in silk drawings unearthed from Mahuangdui in Changsha. Judging by the pottery and lacquer ware unearthed with these papers, it was obvious these were from around 141 BCE (Fan, 1992, p. 299–300).

These archaeological discoveries provide indisputable evidence that paper was already in existence well before CE 105. Tsien's (2004) research of archaeological evidence and records even dated the origin of paper to the third century BCE (p. 145–146). What, then, was Cai Lun's role in the invention of paper?

Cai Lun's role, as Tsien (1962) points out, might have been "that of a sponsor or promoter who, as part of his official duty, reported to the throne the method he had observed" (p. 135). Following is a brief account of Cai Lun (spelt as Ts'ai Lun in the translation) and his invention as described by Fan Ye in *Hou Han Shu* in the fifth century (cited in Carter, 1955):

> During the period Chien-ch'u (CE 76–84), Ts'ai Lun was a eunuch. The emperor Ho, on coming to the throne (CE 89), knowing that Ts'ai Lun was a man full of talent and zeal, appointed him a *chun ch'ang shih*. In this position he did not hesitate to bestow either praise or blame upon His Majesty.
>
> In the ninth year of the period Yung-yuan (CE 97) Ts'ai Lun became *shang fang ling*. Under his instruction workmen made, always with the best of materials, swords and arrows of various sorts, which were models to later generations. (p. 5)

In ancient times writing was generally on bamboo or silk, which were then called *chih* (or *zhi*). But silk being expensive and bamboo heavy, these two materials were not convenient. Then Ts'ai Lun thought of using tree bark, hemp, rags, and fish nets. In the first year of the Yuan-hsing period (CE 105) he made a report to the emperor on the process of papermaking, and received high praise for his ability. From then on paper has been in use everywhere and is called the "paper of Marquis Ts'ai." (p. 5)

The Process of Papermaking in Han

Papermaking techniques and processes were already rather sophisticated in Western Han (206 BCE—CE 25), before Cai Lun's time, and then were further improved in Eastern Han (CE 25–220), when Cai Lun expanded the selection of materials and significantly improved the papermaking process. Fan (1992) has a detailed account of the papermaking method in Western Han, which included a systematic six-step process of material selection, pretreatment, chemical treatment, manual pounding, mixing and distilling, and post-treatment (p. 300–302).

Step 1, Material Selection—All papermaking materials involve plant fiber, but not all plant fibers are appropriate for papermaking. What determines whether a particular fiber is appropriate for papermaking is two factors: its physical property and its chemical property. The physical property refers to the ratio of the fiber's length to its width. The longer and the thinner the fiber is, the better it is for papermaking. This is because longer fibers bind together better. Hemp fibers' length-width ratio can be as high as 3,000, compared with 100 to 200 of bamboo and hay fibers. The chemical property refers to the content of cellulose in the fiber. The higher the cellulose content, the better the fiber for papermaking. In this sense, hemp fibers' cellulose content is as high as 82%, compared with about 40% for bamboo and hay fibers. Analyses of archaeological evidence found that most of the materials selected for papermaking in Western Han were hemp, which meant the paper made then was already rather durable even though it was nowhere as refined as today's paper.

Step 2, Pretreatment—Pretreatment refers mainly to the soaking, cutting, and washing of the materials. Papermaking in Western Han already involved such an essential step.

Step 3, Chemical Treatment—Apart from cellulose, plant fibers also contain some other elements, elements that are, unlike cellulose, undesirable for papermaking. These elements include ash, pectin, protein, pigment, etc., which all affect the final quality of paper. Early papermaking in China already employed various means to eliminate these unwanted elements in plant fibers. As early as pre-Qin times, for example, the Chinese had already discovered the method of using hot water to eliminate pectin from plant fibers. Post-Han Chinese also discovered lime wash for the same purpose.

Step 4, Manual Pounding—In this step, hemp fibers soaked in lime water were placed in stone mortars, manually pounded for thinning the fibers, and then washed to eliminate the unwanted elements.

Step 5, Mixing and Distilling—After pounding and washing, the hemp materials were then soaked in a water groove and then mixed using a wooden stick so that the fibers would spread out evenly and float to the surface to form a pulp, which was then filtered through a sifter to form a thin paper coating.

Step 6, Post-Treatment—This last step involved pressing and drying the paper until it became usable.

By Eastern Han, this papermaking process was evidently further improved. Judging by the finer quality of paper unearthed from this period, the improvement was rather significant. In addition, Cai Lun and the like expanded the material selections to include tree bulk, cloth, and fish net, the very use of which required better, improved papermaking techniques.

In the next few centuries after its invention, papermaking technology was further developed and perfected. Paper became the main medium of writing and gradually replaced bamboo and silk for its obvious advantages: light weight and low cost. It became popular in China and spread to various parts of the world:

> It traveled Westward to Chinese Turkestan in the third century, to Western Asia in the eighth century, to Egypt in the tenth century, and reached Europe in the twelfth century. It also spread eastward to Korea in the fourth century, to Japan in the fifth century, and southward to India before the seventh century and to Indo-China even earlier. (Tsien, 1962, p. 138–139)

The impact of the invention of paper on the world was nothing short of a revolution and has been well researched and elaborated (see Carter, 1955; Hunter, 1947; Jian, 1979; Tsien, 1962). Since my emphasis is less on the significance or impact of the invention than on the context, I will take a look at the rhetorical elements within that context that enabled the invention of papermaking technology.

The Rhetorical Context

To better understand the rhetorical context that facilitated the invention and development of papermaking, it is necessary to give a brief overview of the chronological context encompassing the Qin and Han Dynasties. When Qin Shi Huang (First Emperor Qin) unified China and established the first centralized government in 221 BCE, it marked the end of the Zhou period and the beginning of the Qin Dynasty. Qin, however, lasted for only a short period of time, until 206 BCE, when it was overthrown by peasant uprisings. The ensuing Han Dynasty (206 BCE to CE 220) was divided into two periods: the Western Han Dynasty (206 BCE to CE 25) and the Eastern Han Dynasty (CE 25 to CE 220). The Western Han Dynasty was a period of both peace and turmoil, with the first part being relatively peaceful and prosperous and the second part marked by turmoil in the form of numerous peasant uprisings, especially toward the end of the period (Jian, 1979, p. 103–161). The Eastern Han Dynasty was a relatively peaceful period and went through major developments in many areas of life, though it eventually met with the same destiny as all the other dynasties in Chinese history—demise. Many factors, political, economic, and cultural, in the social context of the Han Dynasty contributed to the invention of papermaking technology in this historical period and to its ensuing development.

Exigency—Calligraphy, the Imperial
College, and Intellectual Studies

Several major cultural developments around the Han period set the stage for the invention of paper. One impetus came from the Qin dynasty. As the first centralized government, the Qin regime took special efforts to unify the empire in many aspects, such as administrative units, currency, and measurements. One such effort in cultural development was the standardization of language. Though characters in the scripts used by different nations more or less resembled one another in their structures, there nevertheless existed many differences. Using the script of Qin as the basis, the Qin government enforced a standardized script as the standard written language of the empire (Jian, 1979, p. 95). This standardization no doubt greatly enhanced cultural and intellectual activities, one of which was calligraphy.

Chinese calligraphy has a history almost as long as that of the script itself. In ancient China, calligraphy was considered a symbol of learning, a sign of intellectual development. Calligraphers were usually held in high esteem. As a result, intellectuals of all periods were almost always calligraphy enthusiasts. The Han intellectuals were certainly no exception. Such a development, according to Hunter (1947), created favorable conditions for the invention of paper:

> The rapid development of calligraphy by archaic Chinese scholars and their spontaneous adoption of the camel's-hair brush and fluid pigment brought forth the exigency for a writing substance that was cheaper and more practical than woven textile. It was this urgent need for a totally new writing surface that inspired the Chinese eunuch Ts'ai Lun CE 105 to proclaim his marvelous invention of true paper—a thin, felted material formed on flat, porous moulds from macerated vegetable fibre. With the advent of paper the art of calligraphy as originally conceived by Ts'ang Chieh in 2700 BCE had its real impetus and the brush-written manner of recording history and setting down accounts upon paper was destined to supersede all other methods. (p. 4)

Calligraphy was certainly not the only impetus that created the exigency. Another major impetus came from the establishment of the Imperial College in the Western Han in the second century BCE The Imperial College was an institution of higher learning established by Emperor Wu Di. In such a school, there were court academicians, similar to the PhD's and professors of today, who taught talented people and trained them to be future officials of the imperial government (Jian, 1979, p. 126). The Imperial College provided a means to an official career for those who otherwise had no way of entering officialdom. Naturally, it became a popular school and developed rapidly, as can be demonstrated by the following account by Huang (1990):

> An imperial university was established by Wu-di, and toward the end of the pre-Christian era, the enrollment had already reached 3,000. When Wang Mang was regent, he reportedly constructed a dormitory that had 10,000 compartments to house an equal number of students, a figure that is more than likely an exaggeration. (p. 49)

The Imperial College had two major impacts. One was that many of the later rulers of the Chinese empire were intellectuals since they graduated from the Imperial College. The notable graduates include Liu Xiu, the founder of the Later Han; Deng Yu, Liu Xiu's adviser; Zhong Chong, the imperial tutor of the heir apparent; and half a dozen top generals (Huang, 1990, p. 50). Another impact of the school was that it promoted learning and scholarship among the people. The fact that at one point about 30,000 students were either enrolled at or affiliated with the school (p. 50) was evidence in itself that learning was a popular activity in the Western Han. In fact, private learning developed simultaneously: "It became fashionable for renowned scholars to gather 500 or more disciples around themselves; the most accomplished had 3,000 followers" (p. 50). Though people of the lower classes were still unable to avail themselves of most opportunities for education, this kind of learning at least was no longer the exclusive privilege of the aristocrats. Such a phenomenon no doubt contributed to the demand for a new writing material that was cheaper and easier to use than the silk.

A third impetus came from the thriving of three kinds of studies: the Confucian classics, Chen Wei—the study of divination combined with the mystical Confucianist belief, and the ancient classics. In both the Western Han and the Eastern Han, many intellectuals took it both as a must and a fashion to study the Confucian classics, namely the Four Books (*The Great Learning, The Doctrine of the Mean, The Analects,* and *The Book of Mencius*) and the Five Classics (*The Book of Songs, The Book of History, The Book of Rites, The Book of Change,* and *The Spring and Autumn Annals*). Along with this interest in the Confucian classics came a popular practice to collect and systematize books, which led to the copying of many books.

Chen Wei started in the late Western Han and prevailed mainly in the Eastern Han. Combining divination with the mystical beliefs from Confucian classics, intellectuals tried to use superstitious beliefs to explain various natural phenomena. As a result, certain natural phenomena were mystified and considered the determining factor of people's lives.

As a counter to this mystic study of Chen Wei and the thriving of Confucian classics, there emerged a revival of interest in the ancient classics. Some scholars believed that people could learn more from the ancient classics. As a result, there emerged many books that were written to explain the ancient classics. These books contributed greatly to later studies of the ancient Chinese script, ancient history, and the ancient classics.

These three factors—the popularization of calligraphy, the establishment of the Imperial College, and the thriving of the three kinds of studies—resulted in the thriving of learning and education, which in turn led to the rhetorical exigency for better writing tools and materials. The invention and development of papermaking in this period, then, was a natural result of social progress.

Ideology—Dominating Confucianism

In terms of ideological thinking, the brief Qin Dynasty, which preceded the Han and lasted for only 15 years, was not one marked with social progress; rather, the autocratic regime enforced by Qin Shi Huang resulted in more regression than progress. The notorious incident of "the burning of books and the persecution of Confucianists," a perfect example of the cruel despotism practiced by Qin Shi Huang and his regime, exerted a reactionary impact on social, especially cultural, development and did nothing but impede any progress in writing technology development. This regression in social progress was not an insignificant factor in the rapid demise of the Qin Dynasty.

In the ensuing Han Dynasty, which lasted more than 400 years, the development of cultural ideologies went through several distinctive stages. In the early Han, immediately after Qin's demise, the lesson of Qin's treatment of intellectualism and cultural ideologies was not lost on the early Han rulers, who took conscious efforts to revive the diverse schools of thought and cultural ideologies that had sprung up during the Warring States period but were stunted in the Qin Dynasty. Various books of the private schools from the pre-Qin era began to re-emerge, as did the intellectuals of the various schools of thought, including the Confucianists, the Taoists, the Legalists, and the Political Strategists, as well as intellectuals of the Yin-Yang School. As diverse an array of cultural ideologies as these schools of thought represented, they served the needs of the early Han rulers, who were looking for a more liberal ideology to guide their regime so that they could avoid the fate that befell the Qin regime.

For the most part of the Han Dynasty, however, what ultimately prevailed as the guiding cultural ideology was Confucianism, or a new version of Confucianism, to be exact. This new version of Confucianism, or what I'd call Neo-Confucianism, was developed by a Han intellectual and official named Dong Zhong Shu (179–104 BCE). Neo-Confucianism differed from the pre-Qin Confucianism in that it incorporated the doctrines of several other schools of thought. Grounded in the Confucianist principle of the unity of God and humans, Neo-Confucianism incorporated the Legalists' principle of centralized governance and the Yin-Yang School's doctrine of the five virtues, but his Neo-Confucianist principles were aimed at rationalizing the Han rulers need to monopolize their ruling and cultural thought. Dong advocated championing Confucianism and undermining all the other schools of thought, but not to the extent of banning all the other schools of thought. Eventually, Dong's Neo-Confucianism established a feudalist political ideology that was to be used for the more than 2,000 years to come (Bu and Zhang, 1992, p. 480–487). Despite this privileging of Confucianism, however, the eclectic nature of Neo-Confucianism fueled, in reality, the various other schools of thought at the time, which in turn provided impetus for intellectual studies. The emergence of Taoism and Buddhism, for example, was a good case in point. However, since both Taoism and Buddhism appeared rather late in the Han Dynasty, their influence on the early developments of paper was less than substantial.

Participants—Rulers and Intellectual Elites

The prosperity of papermaking technology in the Han Dynasty was a result of promotion by two major, powerful groups of participants in its process—the rulers and the intellectual elites. The rulers were responsible for taking the initiative of making significant policies on intellectual studies while the intellectual elites were credited with practicing and promoting the studies.

The Han rulers could take a lot of credit for promoting, although indirectly, the papermaking technology. As mentioned earlier, Dong Zhongshu's advocacy of Confucianism as the official cultural ideology prompted the early Han rulers to adopt a national educational policy squarely grounded in Confucianism. Confucian classics were used as the basis for selecting talented people and officials. As early as 136 BCE, Han established the system of appointing court academicians who were well versed in the five Confucian classics, and by 124 BCE, the court academicians were assigned 50 followers. This court academician system established Confucian classics as the main content of schooling and was considered the beginning of government sponsored official education (Yu, 1992, p. 393–394). Confucian classics, then, were seen as the main path to an official career. In this case, government's role in the promotion of intellectual studies, and thus of papermaking as a means for the studies, was instrumental.

A second group of participants—the intellectual elites—were the direct beneficiaries of the government's policy on education. As a concrete measure of the governmental involvement in the promotion of intellectual studies, the Imperial College was established in 124 BCE, when the Han rulers began to appoint court academicians. Students for the Imperial College came from two main sources: direct selection by a specially appointed court official of those over eighteen, "with good appearance and manners," and recommendations by local government officials of those who "loved literature, respected their elders, valued politics and education, were pleasant to neighbors, and had good behavior" (Yu, 1992, p. 399). The court academicians acted as their teachers, or masters of learning. During the heyday of the Imperial College, students numbered as many as 30,000. The main purpose of their study at the Imperial College was to clear their path to official careers. Their engagement in the study of Confucian classics, and in intellectual studies in general, provided a direct impetus for the use of books and paper. Although it could be argued that these students were not necessarily elites since many of them came from the low social stratum, the fact that most of them would later become government officials after graduating from the Imperial College removed them from their original social stratum and really put them into the elite class.

Also noteworthy in the promotion of official education and thus of papermaking were two other groups of participants, more accurately, however, two subgroups of the ruling class. One was the local governments. As a direct response to the imperial policy on education, local governments followed suit and established official schools at the local level as a source for new government officials. The other group was the imperial family and aristocrats. Recognizing the importance of official education in legitimizing official positions, Han rulers opened special schools for members of the imperial and aristocratic families. The focus of these schools was to provide education to the Emperor himself and heirs to his throne—his children. Whether subjectively or objectively, these groups of participants played pivotal roles in promoting learning, and thus the use of books and papermaking, in the Han

Dynasty. While there's no denying that the general public must have played some role in the development of papermaking technology in Han, the deciding impetus, in my opinion, had come from the ruling class and the intellectual elites.

Knowledge Creation—The Distinct Status of Books

Paper in the Han Dynasty was undoubtedly considered a symbol of status as it was associated with knowledge and learning and often led to official careers. This was due to two factors—the increased availability of paper and the government's emphasis on learning.

As papermaking technology perfected its process over the long duration of the Han Dynasty, especially after Cai Lun's improvement of the technology in the latter part of this period, the Eastern Han, paper, for the first time in Chinese history, became the widespread medium of learning. Not only were the Confucian classics published in book form, but major works appeared in almost all disciplines, such as *Shennong's Herbal Classics,* the first work in pharmaceuticals; *The Classic of 81 Complicated Cases,* a major medical work that complemented *Huangdi's Classic of Internal Medicine,* the first comprehensive work in medicine, completed in the Warring States period; multiple major works in agriculture; *The Book of History,* the first comprehensive work of historical studies; *The Han Chronicles,* the first chronicle of dynastic history; multiple works in religion, such as Taoism and Buddhism; *The Book of Music* and *The Book of Rhythms* in music; and numerous works in prose, poetry, and folklore, such as the well-known *Peacocks Flying to Southeast.* The emergence and publication of multiple books of significant influence in almost every single field in the Han Dynasty marked a milestone in the establishment of books' status as the authoritative medium of knowledge.

A second contributing factor for the high status of books was the Han rulers' emphasis on learning. As education became the only major means to official careers, the study of written classics, especially the Confucian classics, rendered the book medium an essential tool in people's quest for political elitism and distinction. This emphasis on learning, coupled with the flourishing of book publication, effectively transformed paper into a symbol of a knowledge-bearing medium of vast significance.

Knowledge Access and Control—Limited Parameters

From the preceding Zhou Dynasty through the current historical period in discussion, Qin and Han, access to paper and books took some twists and turns. The Qin Dynasty represented an era of regression in this respect as First Emperor Qin's anti-Confucianism campaign resulted in the burning of books and the persecution of intellectuals in a desperate attempt to rein in cultural ideology and political thought. Access by the general public to books and paper, as a result, dwindled to diminutive levels.

The Han rulers took a different approach by encouraging intellectual studies, with a particular bias towards Confucian classics. However, access to paper and books in the Han period was not exactly one with wide-open parameters. Instead, the Han rulers were caught in a dilemma: on the one hand, lessons from the Qin Dynasty prompted them to take a more liberal approach towards public's access to this technology and the knowledge associated with it; one the other hand, they were also wary of the potential threat too much access to knowledge by the populace could pose to their reign. What resulted from this paradoxical attitude was limited parameters in both control by the ruling class and access by the public.

Limited control of access to paper and books was a result of the Han government's open promotion of learning and education. As the Imperial College at the central government level and official schools at the local level were established as a direct measure of the government's policy on learning and education, private schools and education re-emerged to fill the void where official schools failed to meet the demand. Private education during this period included three main areas. One was the counterpart of official education, where adults followed particular masters of learning in various fields to learn the knowledge and crafts. A second main area was early childhood education, where young children learned either in special schools or from private tutors in their own homes. A third area was home schooling, where both children and adults learned from their elders. An essential part of all this education was the learning of reading and writing. Books and paper then were undoubtedly accessible by the general public. Such a rather open access rendered the Han rulers' control rather difficult.

At the same time, however, this seemingly open access to books, paper, and knowledge enjoyed by the public was not without its limits. Even though the populace could access books, knowledge, and education, those educated in the private schools or at home did not possess the means to turn their knowledge into resources that could lead to official careers since only those who went to the Imperial College or local official schools had the opportunity to go into official careers. Therefore, even though everybody who had education and access to books and paper in either official or private schools had the same access to the knowledge that books and paper represented, they didn't have equal access to the distinction and status that official education could provide. In this sense, access by the general public to books and paper, and especially to what they represented, was limited in many senses.

Communication Medium—Increased Written Dissemination

Official and private schools in the Han Dynasty carried the main load of knowledge dissemination. However, since printing was not yet invented then, the dissemination of book knowledge took two main forms: oral communication and hand copying. Hand copying was time consuming, so oral teaching by the masters of learning formed the bulk of knowledge communication in the Han period. In addition, the Hans believed the correct understanding of the classics must rely on the interpretations of the masters and self-study was not the way for great learning.

At the same time, the written form of communication gained significant ground as the number of books increased significantly and the written language itself matured by considerable measures. Compared with the Zhou Dynasty, a much greater number of books covering more disciplines emerged in the Han Dynasty. Almost every discipline saw books of significant impact appear in this period, as mentioned earlier in this chapter. More importantly, the written form of language went through significant developments and became more complete. First, there was the standardization of the written language in the Qin Dynasty. Second, character styles became more varied after several distinctive stages of development. In the Qin Dynasty, seal characters were used as the official writing style for court records and all formal occasions. By the Han Dynasty, as more writing was required for official records and communication, the more complex and difficult-to-write seal character script gave way to the official script style, a simplified form of seal characters. This development represented the transformation of Chinese script from pictographic representation to abstract depiction. Also developed in Han were the cursive hand and the running hand, two less formal styles more widely used on informal occasions.

A more important factor concerning written language in promoting and communicating about the use of paper and books was the formal research and study conducted of the written script itself. Among many significant books on language published during the Han period, there was Xu Shen's *Shuo Wen Jie Zi* (*Annotated Dictionary of the Chinese Language*), the first encyclopedic dictionary that annotated 10,516 characters in terms of their shape, meaning, and pronunciation (Wang, Li, & Zhu, 1992, p. 419–420). There was *Er Ya* (*Standard Chinese*), the first comprehensive work on critical interpretations of ancient texts. There was also *Dialects,* the first study of ancient and Han dialects. Then there was *Shi Ming* (*Interpretation of Names*), the first systematic study on etymology. In addition, the Han Dynasty also saw the birth of the first significant works of book taxonomy, including *The Book of Han—The History of Art and Language.*

That so many influential works on language studies appeared in the Han Dynasty was a good indication of the significant status attributed to the written language in this historical period. At the same time, what they communicated about the significance of books had a shaping impact on people's perception of paper. As an important and essential complement to the oral medium, the written medium of books served as an effective illustration of the importance of the very medium itself—the paper.

7 Block Printing and Movable Type

> In invention, what the T'ang period conceived, the Sung [Song] era put to practical use. The magnetic needle, used in the main in earlier times either as a toy or for the location of graves, was applied to navigation. Gunpowder, already known and used for fireworks, was during the Sung Dynasty applied to war. Porcelain was so developed as to become an article of export to Syria and Egypt. (Carter, 1925, p. 55)

Although the magnetic needle, gunpowder, and porcelain had nothing to do with writing technologies, the kind of relationship Carter is talking about here between the Tang (618 CE–907 CE) and Song (960 CE–1279 CE) Dynasties in terms of inventions and development is characteristic of the two major printing technologies from this period—block printing and movable type. Block printing, which emerged in the Tang period but never really gained much momentum, paved the way for the more important invention—movable type—in the Song Dynasty.

BLOCK PRINTING

The exact starting date for printing is a matter of controversy. Part of the reason for this difficulty in determining the date lies in the fact that "the evolution of the art was so gradual as to be almost imperceptible" (Carter, 1955, p. 41). In the long period of time before block printing emerged, the Chinese had already taken a series of steps that led to block print: rubbing from stone, printed silk, stencil, seal, and stamp. This gradual, multi-stage evolution made it impossible to pinpoint the origination date for block printing. However, some researchers seem to agree that the first extant example of block printing dates from the million printed *dharani* of the Empress Shotoku of Japan of 770 CE (see, for example, Carter, 1955; Hunter, 1947; Twitchett, 1983; *Common Facts*, 1980). The Japanese technique is believed to have been taken from China (Twitchett, p. 14). As for the next century, there is "abundant literary evidence that block printing was being practiced in China" (Twitchett, p. 14).

Other researchers tend to disagree. In his *A New History of China* (Volume 2), Bonian Guo (1984) sets the origination date for block printing in the Eastern Jin period (317 CE–420 CE). According to Guo's account, the priests in Eastern Jin often had to deal with the burdensome task of writing and drawing their magical incantations each time they performed rites to help people cure diseases and dispel misfortunes. To make the task easier, they carved the incantations on jujube wood blocks so that they could print it each time such incantations were needed. This invention was obviously inspired by the use of stamps in the Qin period. However, these incantation blocks were often much bigger than the Qin stamps, and some of them held as many as 120 characters. This, according to Guo, was really the embryonic form of block printing, and by Tang, this technology had already matured (p. 515).

Using archaeological evidence, *Common Facts* (1980) seems to corroborate Guo's account at least to the extent that the block printing technology had matured by the Tang period. According to *Common Facts*, a printed copy of *The Buddhist Scripture*, dating to 868 A.D, was found in Dunhuang in 1900. It consisted of seven printed pages attached to one another and was 533 centimeters long. This was believed to be the first printed material ever found in the world. The Buddhist Scripture featured superb engraving and fine printing, providing proof that the block printing technology had been perfected to quite some extent by the Tang era (p. 182).

Carter's description in his 1955 study of block printing as it is practiced today gives us some idea of what block printing in China means and has meant for the past couple of thousand years or so. According to Carter, the printing block is usually made of pear or jujube wood, which is finely planed and smoothed in the size of two pages. The pages to be printed are first carefully transcribed onto thin sheets of paper by a professional. The sheets are then pasted onto the wood block in a reversed position. When the sheets are subsequently rubbed off, the impression of the characters in ink remains on the wood block.

Then a workman uses a sharp graver to cut away all the portion of the wooden surface not covered by the ink. After the block is cut, the printer uses a brush to apply ink onto the surface of the characters. Then with paper laid on the block, he uses another brush, a dry one, to run over the paper to take the impression. "The paper, being so thin and transparent, is printed on one side only and each printed sheet (consisting of two pages) is folded back, so as to bring the blank sides in inward contact. The fold is thus at the outer edge of the book and the sheets are stitched together at the other" (Carter, p. 34–35).

This form of printing was later improved by Feng Dao (882–954), the prime minister of the later part of the Tang dynasty (618–907 CE). Though Feng Dao was often inappropriately credited with the invention of block printing, he did play a significant role in improving and popularizing the art. By the Song (Sung) Dynasty (960–1279 CE), block printing had already become such a refined art that "the books of the Sung Dynasty have never been surpassed in printing skill" and "Chinese books printed from modern type cannot compare with them" (Carter, 1955, p. 32).

Curiously, and unfortunately, the development and use of block printing arguably never really gained the kind of momentum it deserved. As mature as the technology was by the late Tang Dynasty and as perfected as it was by the Song Dynasty, block printing, though a notable invention, has a rather dubious reputation in the history of Chinese culture, not so much because it wasn't significant as because it never exerted the kind of impact other writing technology inventions, such as paper, have on the evolution of the Chinese civilization. Tsien (2004) explains it this way:

> The increasing demand for more religious literature stimulated the development of printing. This new application, however, did not end the production of books by hand. Nor did it change the general nature of their format, content, quality, or even quantity in creative writing. What it did was merely to increase the number of copies by facilitating duplication of the writing, so that communication by books became wider and easier. Nevertheless, quantity did not necessarily improve the quality or the depth of the writings. Even today, mass media had not produced a book that can compare with the originality or creativeness

> in style and language such as the *Shiji* (*Records of the Grand Historian*), written on hundreds of thousands of bamboo tablets by Sima Qian more than a thousand years ago and that has been followed by all the dynastic histories ever since. Also, human nature once again demonstrated its conservatism, for printing did not become popular until some three hundred years after its introduction in the seventh or early eighth century. Commemorative writings have always been inscribed on permanent materials, and manuscripts have since been more treasured than printed editions even down to our times. (p. 205–06)

One could easily turn to the historical context of block printing technology to look for clues. However, besides what Tsien has discussed above, one important factor was that the complexity, the cumbersomeness, and the time-consuming aspect of block printing never really rendered this technology an easy and practical application. Even though printing became popular some three hundred years later, in the Song Dynasty, as Tsien has mentioned, the printing that Tsien is referring to here is not block printing, but rather movable type.

Movable Type

Movable type was born as a logical solution to the problems of block printing. With block printing, it often took years and years to produce a substantive amount of printing blocks, but most of these blocks were used only once. Not only was this a huge waste, but the storage of these mountains of blocks also posed real problems. Such a wasteful practice led to the high cost of books and was a serious hindrance to book printing and cultural dissemination. By the Song Dynasty, this printing technology, though still in use, apparently fell short of accommodating the high demand for written communication. Movable type, with its flexibility of character selection, arrangement, and reuse, effectively filled this gap in printing technology.

Movable type was invented around 1041–1048 in the Song Dynasty (960–1279) by a man named Bi Sheng (also spelled as Pi Sheng). According to Xia et al. (1979), Bi Sheng used sticky clay and cut characters in it, with one character in each piece of clay. He then baked the characters in the fire to make them hard. Then on an iron plate with

Block Printing and Movable Type

frames, he put a layer of pine resin wax mixed with paper ashes. After placing the clay characters in the frames, he put the iron plate over the fire. When the wax was heated to a certain degree and melted a little, he used a smooth board and pressed it on the surface of the type so that the block of type became even. Once the plate cooled down and the wax became hard again, the type became fixed on the plate. Then the type was ready to be used for printing just like block printing. Later, wood and metal types were also developed based on Bi Sheng's clay type.

The advantage of movable type printing over block printing lies in its ability to produce a large number of copies in a small amount of time. If it is used to duplicate something with only a few copies, it does not really have any advantage over block printing. However, with large quantities of printing, block printing cannot be compared with movable type. The advantage of the latter in mass printing probably lies in the way printing was done. As Shen Kuo, a contemporary writer of Bi Sheng's time, explains:

> As a rule, [the printer] kept two forms going. While the impression was being made from the one form, the types were being put in place on the other. When the printing of the one form was finished, the other was all ready. In this way the two forms alternated and the printing was done with great rapidity. (Carter, 1955, p. 212)

Nevertheless, this superior printing method did not become popular until Gutenberg reinvented movable type in the mid-fifteenth century, a whole four hundred years after Bi Sheng's invention.

THE RHETORICAL CONTEXT

Since block printing and movable type emerged in two distinctive periods that were three hundred years apart, each calls for separate examinations of their respective contexts, the Tang Dynasty for block printing and Song for movable type. Even though the origination date for block printing is uncertain, and even though the use of block printing never saw its heyday in the Tang Dynasty, the fact that this printing technology became rather mature in the Tang period justifies at least a brief look at its rhetorical context.

That the Tang dynasty saw the invention of (block) printing, traditionally considered, even in Chinese textbooks, to be one of the four greatest inventions of China (the other three being paper, gunpowder, and the compass), is no accident, if we consider the historical and social circumstances of the time. The nearly three centuries of the Tang dynasty marked a period of high stability, one that was, relatively speaking, probably the most prosperous of all periods, or, in Carter's (1955) words, "one of the most glorious in the history of China" (p. 37). In addition to political stability, military prowess, and economic prosperity, the Tang period also witnessed great spurts in literary and intellectual development. The literature, especially poetry, from this period represents some of the highest attainments of all times. The Tang culture nurtured such great poets as Wang Han, Wang Zhi Huan, Wang Chang Ling, Meng Hao Ran, Wang Wei, Li Bai, Du Fu, Bai Ju Yi, Du Mu, Li Shang Yin, etc. (Jian, 1979, p. 234–239). In addition, painting and calligraphy were also thriving. These literary and intellectual developments created some of the most favorable conditions, and a big demand, for a duplication method that would be easier than hand copying.

Perhaps, as was the case with most of the other inventions in writing technologies in the history of China, a greater factor shaping the invention of block printing was the attitude of the Tang rulers. As Carter (1955) notes, "the early emperors of the T'ang Dynasty were great patrons of literature, of art, and of religion" (p. 37). For example, a library housing over fifty-four thousand rolls was built under Emperor Tai Zong (also spelled as Tai Tsung, 627–649), whose impartiality in religious toleration has also "seldom been surpassed in history" (Carter, p. 37). As a result, partly, of the natural development of things by that historical point and, partly, of the Tang emperor's open mind, different ideological schools of thought, religions, and even the first Christian missionaries thrived in that period. In that context, printing was not only a medium for writing, for communication, and for artistic expressions, but also a tool for advancing the educational and intellectual development of the empire, its other fields such as science and technology, and consequently the rule of the emperor. As Twitchett (1983) points out, "The state . . . very quickly appreciated that printing could not only make books widely available; it could also be used as means of state control and of ensuring conformity" (p. 32). Thus, in a totalitarian society such as that of China, the attitude of the ruler can often be the key factor in determining the fate of a technological innovation.

Nevertheless, as noted earlier, block printing was never popularized due to the complex nature and impracticality of this technology. Its perfection, however, in the Song period and, more importantly, the critical role it played in facilitating the invention of movable type tendered historic significance for this un-popularized but nevertheless milestone invention. It is perhaps more meaningful to look at block printing as an early stage in or a precursor to the invention of movable type, the invention that had much greater far-reaching influences both in and outside China. My examination of the rhetorical context for the invention of printing technologies, therefore, will focus on the Song period, when movable type came into existence.

Exigency—Cultural Prosperity and Scientific Development

What exactly prompted the invention of movable type is not totally clear. However, a couple of major factors in the social context of the Song Dynasty at least contributed to this invention. First, printing thrived in this period. Though the Song Dynasty was not one of great military strength and political stability, it was "one of the great peaks of Chinese culture, particularly in the visual arts" (Twitchett, 1983, p. 34). This cultural prosperity was triggered partly by the printing of the Confucian classics under Feng Dao's administration in 953, seven years before the Song Dynasty began (Carter, 1955, p. 211). This production marked the inauguration of the era of large-scale official and secular printing" (p. 211). This large-scale printing included "massive state printing projects aimed at the production of standard editions of orthodox literature," the issuing of "a stream of further commentaries on the Confucian canon, encyclopedias, literary anthologies and dictionaries" by the National Academy, and the publishing of "a standard printed edition of the seventeen Standard Histories of all earlier Chinese dynasties, a massive undertaking that took from 994 to 1063 to accomplish" (Twitchett, p. 34).

This "progress . . . in the realm of printing" according to Carter (1955), "was itself one of the causes of the spirit of advance" (p. 211). The rapid printing of all sorts of books naturally urged men to look for improved printing methods. According to Carter, this revival of interest in Confucian literature and the ensuing interest in printing methods resulted in various experiments in printing methods (p. 211).

Many such experiments were in block printing, including the use of blocks made of materials other than wood. As a result of the increase in printing and improvements in printing methods, printing in the Song dynasty reached quite a height. As Twitchett attests,

> Although never approaching the dazzling technical virtuosity of late Ming and early Ch'ing block printers with their use of multiple impressions, multi-coloured printing, etc., the style, calligraphy and technique of Sung printers set a standard constantly imitated in later times, but never surpassed for their beauty. 'Sung style' was a claim made for their better wares by block printers in later centuries, and even today one of the most common Chinese typefaces is called 'Old Sung.' (p. 34)

With such a context, Bi Sheng's invention of movable type was more a natural result, in the wake of this thriving interest in the improvement of printing methods, than a surprise development.

A second major factor that contributed to the invention of movable type was the scientific development in the Song period. In the Song dynasty, as Carter (1955) points out, "in spite of political upheaval and financial chaos, especially in the dark days of the thirteenth century, science and philosophy went forward" (p. 211). Song was a period of great scientific development. Of the four greatest inventions in the history of China, namely, paper, printing, gunpowder, and the compass, two (gunpowder and the compass) were invented and one (printing) improved/reinvented all in the Song dynasty. Other, less prominent inventions in many areas also emerged during this period. It seems that the overall culture of the Song dynasty was conducive to technological inventions.

Ideology—Self Assertion vs. Social Conformity

In the case of all the writing technologies prior to movable type, the inventors all held certain official positions in the government. Unlike his predecessors, Bi Sheng, the inventor of movable type, could claim no official affiliation whatsoever. Movable type marked the first invention by the common people. Though this no doubt can be attributed to various factors, the political and cultural ideology during the Song period played a more-than-negligible role.

That Bi Sheng, a commoner with no official affiliation, could claim such an important status in the invention of the movable type could be attributed, at least in part, to a notable distinction about the Song Dynasty—the transforming perspectives in political ideology. In feudal societies such as Song and several preceding dynasties, the emperor and his royal sovereignty had always been held in the utmost regard and were allowed the absolute power to determine the fate of everything, from policy making to the life and death of his citizens. So, like in previous dynasties, the driving force for writing technology inventions in Song could very well have come from the monarch (and part of it, indeed, did). However, beginning in the early part of Song, this concept of monarchial autocracy began to be challenged by such political reformists as Fan Zhongyan and Wang Anshi, both of whom advocated a more rational approach to political rule based on morals and talent instead of autocratic force. Such political reformation was more or less adopted by the Song rulers and facilitated a system of more limited monarchial power and less centralized imperial sovereignty. This decentralization of political power, although limited, made it possible for commoners like Bi Sheng to assume and play more significant roles in technological innovations.

According to Bu and Zhang (1994), this reformation in political ideologies, together with changes in economic and class structures, inevitably brought about changes in intellectual and cultural ideologies, rendering possible the pursuit by a large number of scholars and intellectuals of cultural and intellectual studies (p. 485). This widespread pursuit of cultural and intellectual studies provided a great impetus for the exchange of intellectual thoughts, which in turn spurred schooling and education.

By the Song dynasty the Chinese civilization had developed to a rather mature point where education was fairly widespread. As Carter (1955) recounts:

> While the T'ang dynasty (618–907) has a freshness that reminds one of the Elizabethan Age, the time of the Sung [Song] emperors (960–1279) has a polish, a love of system, and a scientific spirit that are essentially Victorian. The Sung mind was that of the modern man. Accounts written at this time concerning the early history of the human race have about them an evolutionary flavor seldom met with in Europe before

the last century, while the financial and social reforms of Wang An-shih would have pleased the liberals of Victoria's day. (p. 211)

This kind of rather enlightened social culture in the Song period led to a cultural ideology that seemed to value intellectual development. That also explains why "in spite of their military weakness, the Sung kept up a high and even luxurious culture in terms of philosophy, literature, and art" (Lattimore, 1946, p. 63). Due to this prosperity in cultural development, the use of writing technologies was obviously no longer an exclusive privilege of the ruling class or the high brow. As writing materials and technologies became more accessible to the common people, men of no distinct heritage began to avail themselves of the chances to get involved in technological innovations.

Participant—"The Common People"

In each of the cases of writing technology innovations we have examined, we see that, in addition to all the necessary social elements that created a strong exigency for the particular writing technology at issue, the technological innovation always required a powerful agent, such as the emperor. Such an agent, it seems, is an essential element to enable the technological innovation in a totalitarian society like China. As we have seen, in the cases of oracle inscriptions, paper, and block printing, the rulers, usually emperors, were an active force in initiating or furthering technological innovations by either directly involving themselves in the innovations or appointing high government officials in charge of the innovation projects.

In the case of movable type, however, there was a seeming lack of active involvement by the rulers or government officials in the invention process. Bi Sheng, who invented movable type, was nothing but an ordinary man. He was neither a government official nor a prominent scholar. Shen Kuo (1030–1094), a contemporary of Bi Sheng and a well known scholar who wrote *Meng Xi Bi Tan* (*Dream Pool Jottings*), which recorded the inventions of the compass and movable type, called Bi Sheng "a man of the common people" (cited in Carter, 1955, p. 212). Undeniably, Bi Sheng's role in this invention was instrumental.

At the same time, dismissing the government's role in the invention of the movable type as non-existent would be unfair. Though there lacked evidence of its direct involvement in the invention, its indirect facilitation of the invention was reflected in its cultural and educational policies. Though a totalitarian regime just like any other dynasties in the history of China, the Song government adopted relatively liberal policies concerning education. These policies led to two things that had a direct impact on the invention. One was open access to education opportunities by the populace, which encouraged the involvement of the common people in various inventions and their advancement. The other was the government's advocacy of intellectual studies, especially the Confucian classics, and the high printing demand it spurred. Without such favorable policies to create the kind of exigencies needed for technological inventions, the invention of movable type might very well have taken a different course.

Knowledge Creation—Agriculture over Commerce and Communal Conformity over Self Assertion

An interesting aspect about the invention of movable type is people's perceptions about its impact on the evolution of the Chinese civilization. Some researchers have argued that movable type had a minimal impact on Chinese society, ancient or modern. Important as it has been to the rest of the world, it's true that movable type never really became popular during the Song Dynasty, or during the greater part of the Chinese history that ensued, and has remained, as Carter (1955) has called it, "an unimportant later addition" (p. 31). Logically, movable type seemed a good fit for the Chinese language. As Twitchett (1983) points out,

> Superficially, the use of movable type for printing Chinese was a logical and straightforward development. In Chinese calligraphy and typography each character is considered to occupy a square space of equal size, the characters being arranged in vertical columns, usually separated in traditional printing by a fine line. The Chinese compositor was thus faced with none of the complexities of spacing and layout that confront the Western typographer. (p. 74)

Given this compatibility between movable type and the Chinese script, the failure of this printing invention to become popular, indeed, seems puzzling. Many researchers have attempted to explain this bewildering phenomenon. Carter attributes it to the unique characteristics of the Chinese language and the "Chinese love of calligraphy as a fine art" (p. 32). Hunter (1947) believes that the invention found little use in China because "the Chinese language with its myriad of characters did not lend itself to the use of movable type" (p. 472). Twitchett explains it in much the same way:

> The basic problem in Chinese typography was, and still remains, the fact that the repertory of Chinese characters is virtually limitless. Even today, after decades of efforts at limiting the number of characters in use, a Chinese printer needs an active stock of more than 8000 characters, while newspapers still use about 5000 common characters. Even with such a large stock, rare characters are often required, especially for proper names, and type needs to be specially cut for them just as Pi Sheng had special additional characters made as he needed them. The largest Chinese dictionaries contain more than 40,000 different characters, although many of these are simply variant forms, and none is exhaustive. No Chinese printer ever had a 'complete' font including every Chinese character. (p. 76)

This peculiarity about the Chinese language no doubt accounts in part for the failure of movable type to become more popular and widespread than it has been, especially in the Song dynasty. However, to attribute this failure entirely to the characteristics of the Chinese language would render us, at most, a partial perspective on the real reason for the unpopularity of movable type printing. Other factors have at least contributed to such a result.

A main factor lies in people's perceptions of this particular technology, which were shaped by the particular cultural ideologies of the time, and the meanings and significance they ascribed to the invention. Huang's (1990) account of why movable type and other technological inventions such as the compass, the astronomical clock, sea navigation technologies, and new weapon technologies utilizing gunpowder failed to make their day in the Song dynasty may give us some idea. According to Huang, the lack of systematic follow-up of these technological inventions can be attributed to the fact that the influence of commerce, which in Western Europe has been the jumping-off point for technological breakthroughs, was never able to outweigh that of agriculture in early modern China. The prioritizing of the agrarian culture resulted in an emphasis "on quantity over quality, on homogeneity over diversity, on endurance over ephemeral flashes of brilliance. These conditions worked against further efforts to widen the uses of inventions" (p. 133).

In addition to the lack of a commercial revolution, the abundance of cheap labor was also a contributing factor. As a result, "while the people of the Song showed ingenuity at solving technological problems, they as a rule remained uninterested in seeking labor-saving devices" (Huang, 1990, p. 133).

Another, less apparent, factor lies in what Huang (1990) calls "a special kind of socio-psychology" prevailing in the Song period (p. 134)—the contention and the eventual balance between self assertion and communal conformity demonstrated by Song scholars. On the one hand, their Neo-Confucianist philosophy, which combined Confucian ethics with Taoist and Buddhist metaphysics, allowed them to break away from the Han aesthetic approach and to exercise spiritual self-assertion; on the other hand, their emphasis on self-cultivation also cultivated an inclination to "remain comfortable with the traditional communal living" (Huang, p. 134). The result of this contention between self assertion and social conformity was a delicate balance between the two, with the balance tilting more toward social conformity. This emphasis on social conformity prevented Song scholars from achieving ideological breakthroughs in their thinking, which did not prove to be conducive to technological innovations.

This emphasis on homogeneity, endurance, and social conformity over diversity, ephemerality, and self assertion, Huang has pointed out, is by no means a unique feature of the Song dynasty. In fact, it has characterized most of the history of Chinese civilization. As is explained earlier, the most influential ideologies in China, such as Confucianism, Taoism, Buddhism, etc., despite their differences, all seem to underscore such an emphasis, which symbolizes, at least in part, the orthodoxy of Chinese thought . Technological inventions were more a representation of breakaways from than conformity to this traditional way of thinking.

This may seem as much an ideological issue as one of knowledge construction. However, it was these very ideological inclinations that had largely determined the Song people's perceptions about the meaning, use, and significance of movable type.

Knowledge Access and Control—A Moderate Balance

As is characteristic of most totalitarian regimes, the Song rulers struck a moderate balance between control and access when it came to knowledge and education. The prosperity of intellectual studies during the Song period was largely a result of Song's cultural and educational policies. Inheriting a state of turmoil and destruction from the preceding Five Dynasties (907 CE–960 CE), the early Song regime was in dire need of a cultural renaissance as well as political stability. The cultural and educational policies were aimed mainly at revitalizing intellectual studies under moderate government control.

These policies exhibited four distinct emphases. First, there was the revaluation of Confucius and Confucianism. A significant number of Confucius temples were restored and rebuilt all over the country under government decrees. Confucius was again esteemed as the master of learning. Second, there was the revival of the study of Confucian classics, and the Song government made deliberate efforts at unifying the interpretation of these classics. As a result, many of these classics were re-edited and reprinted. Third, civil officials were placed in important government positions. The government took specific measures to improve the imperial exam system they inherited from the Sui Dynasty to encourage intellectual studies. Fourth, the government facilitated and sponsored large-scale book collection and studies. For example, the Imperial College, the highest educational administration in feudal China, became a critical book printing and collection agency. As a case in point, the collection of books in the Imperial College grew from about 4,000 volumes at the beginning of the Song Dynasty (960 CE) to over 100,000 volumes in 1005 CE, an increase of 25 times over a mere 45 years (Qiao, 1994, p. 441).

What these cultural and educational policies effected was a boom in education, schooling, intellectual studies, and book printing, among a myriad of advancements in other areas. Through this exposure to education, books, and knowledge, the intellectual and the general public had open access to knowledge about book printing and the printing technologies of earlier times. It was no surprise then that a common, average man like Bi Sheng had access to earlier printing technologies and was able to improve upon them with his invention. In contrast, it's not clear how much communication and what kind of communication occurred about the invention of this new printing technology. As revolutionary as this technology might have been, it never garnered the kind of attention it deserved during the Song era. What's more intriguing is that there's no evidence that suggests any deliberate attempt on the part of the Song rulers to control people's access to this new technology. It probably had more to do with the cultural ideologies of the time, as explained earlier.

Communication Medium—Westward Spread

The medium of communication about printing technologies remains, somewhat, a mystery. There are no written records from the Song Dynasty documenting communication about block printing and movable type. What is known for a fact is that both of these printing technologies spread to other ethnic groups and regions within China and to other countries outside China. According to Guo (1984), printing technologies were first spread to Korea, and then to Japan (p. 518). There's also evidence that the Silk Road, a commerce route between Ancient China and Europe, played a significant role in spreading these printing technologies to Europe via the Persians and the Arabs (p. 518). Carter's (1925) book offers elaborate accounts of the spread of printing technologies from China Westward through the Silk Road (or what he calls the Great Silk Way): the printing of Uigur Turks in the region of Turfan, the Mongol Empire as the connection between China and Europe, Persia as the crossroads between the East and the West, block printing in Egypt during the period of the crusades, block printing in Europe in the fourteenth century, and eventually Gutenberg's invention of printing in the fifteenth century, the pedigree of which, according to Carter, can be traced to the printing technologies in China (p. 180–182).

This Westward communication, according to Carter (1925), was realized through five different mediums. The first was through paper, which was invented in China, transmitted through the Islamic world, and used as the foundation for printing. The second was through playing cards, which were also introduced to Europe from China during the fourteenth century. The third was through image prints, most of which, though of European design, suggest Asian creation "in subject matter and purpose, in ink and in technique" (p. 183). The fourth was "through the great number of books printed in China" and the reports of such printing by Europeans returning from China. The fifth, a possibility but not a probability according to Carter, was through reports to Europe of "the actual method of typography in use in the Far East" (p. 183). By this account, it is obvious that these five factors acted both as the mediums of communication Westward about China's printing technologies and as the stimuli for Gutenberg's invention.

8 The Chinese Typewriter

> The inherent problems of typesetting in a script with such a large and almost unlimited number of characters, which finally drove movable-type printing from the market in favour of the wood-block printed book, remain with us today. Even in the age of the computer and advanced photo-setting machinery, the problems of storing and efficiently accessing such an enormous font remain difficult of solution. Perhaps in a parallel to the history of printing techniques, the Chinese have preferred to use in telecommunications the relatively expensive technique of photo-facsimile transmission (the modern-day equivalent of the block print) rather than attempt to store and retrieve the whole repertory of characters in digitized form (the space-age equivalent of movable type). The problem lay not in any failing of ingenuity or printing technology, but in the fundamental nature of he Chinese written language. (Twitchett, 1983, p. 86).

The problems Twitchett is describing here about movable type were the same problems confronting the Chinese typewriter. Of course, what Twitchett obviously couldn't predict when he wrote his book in the early 1980s was that China would develop the first computer with the Chinese script but the same Western keyboard as early as 1985, effectively solving the problem of digitized storage of thousands of Chinese characters which could be easily retrieved by various key combinations. This digital solution, however, was not an option during the invention of the Chinese typewriter.

Perhaps, nothing parallels the Chinese typewriter as the oxymoron of writing technology development in the modern history of any civilization, oxymoronic in that this technological innovation was significant and revolutionary, at least in theory, in the history of writing technology development in China while being trivial and insignificant almost to the degree of total anonymity among the populace. It was significant in that it represented a revolutionary technology that had the potential of finally liberating the Chinese from the pen and the paper in their everyday communication. It was alarmingly trivial and insignificant in the sense that it was never popularized, that the majority of the Chinese populace have never even seen a Chinese typewriter in their lifetime, let alone used one, and that there has been nothing but scarce mention of this writing technology in research on the cultural, scientific, or technological history of modern or contemporary China.

That a simple writing technology could embody such polar opposites when it comes to its place in the history of writing (and of culture) is in itself a topic worth exploration. In a culture where writing has almost always held a significant place and demanded utmost respect from the uneducated populace as well as the highly educated intellectuals, such a writing technology, one would think, should logically have met with great enthusiasm and immediate popularity instead of near total oblivion.

What makes the Chinese typewriter an even more worthwhile topic is the fact that its subsequent replacement technology—the computer—encountered a very different fate and went through a totally different development path, one of immediate acceptance, rapid growth, and surprising popularity. What has accounted for such a huge discrepancy in the fates these two writing technologies has largely escaped the attention of researchers in most fields. More importantly, few have ever attempted to link these two technologies in any meaningful manner or wonder why the Chinese typewriter became a missing link in the history of writing technology development in China and whether and how this missing link contributed to the subsequent development of computer technology as a new writing medium.

The History of Its Invention

The Chinese typewriter was invented by Lin Yutang (1895–1976), a Chinese living in the U.S., around the 1940s, a whole eighty years after the birth of its Western counterpart. A linguist and writer, Lin obtained his bachelor's degree from Saint John's University in China, a Master's degree in Literature from Harvard, and a PhD degree in Linguistics from Leipzig University in Germany. He spent more than thirty years of his life in the U.S. and published a number of well known works in linguistics as well as novels and essays in both English and Chinese. His work in fiction, including *Moment in Peking*, earned him a nomination for the Nobel Prize for Literature (Wang, Z.).

Besides being an established writer and linguist, Lin was also an invention enthusiast and devoted much of his life and financial resources to the invention of the Chinese typewriter. As recorded in "Lin Yutang's Invention of the Chinese Typewriter," as early as 1916, Lin became interested in the idea of a Chinese typewriter and a Chinese character arrangement system. His adventure into this virgin area of invention began with his careful examination of Western typewriters. He dreamed of designing a Chinese typewriter that could produce an unlimited number of characters with a limited number of keys, just like the Western typewriter. After thirty-some years of research, Lin invented the Instant Index System, which was based on the shape of the radicals rather than the order of the strokes ("Lin Yutang's Invention"). It was a rather significant invention as it represented a serious departure from the traditional indexing systems in Chinese.

Lin's invention process for the Chinese typewriter, however, was filled with countless complexities in the design and making of the parts, the finding and hiring of expert engineers, the making of the typeface, the arrangement of characters, and the incredible amount of time and financial resources required. On May 22, 1947, after more than 30 years of Lin's persistent efforts and more than $120,000 of Lin's financial investment, the first Chinese typewriter was finally born. Lin's second daughter, Lin Taiyi, did the honor of the first hunt and peck and successfully produced Chinese characters on a piece of paper with this typewriter.

The marketing of this first Chinese typewriter was no less frustrating than its invention. When news of Lin's invention got out, Remington, the company that manufactured the first Western typewriter in 1873 (Beeching, 1974, p. 88), showed interest in Lin's invention and asked him to bring his invention to their company for a demonstration. Unfortunately, with a dozen senior Remington officers watching, Lin and his daughter couldn't get the typewriter to work, and their demonstration turned into a complete disaster. A disappointed Lin went home and summoned the Italian engineer who designed most of the parts for the typewriter. The Italian engineer was able to identify and fix the minor problem that had prevented the typewriter from working properly. By that time, however, it was too late to reverse the negative impression the failed demonstration had left on Remington's management.

The next day, the pre-scheduled press conference for the invention of the Chinese typewriter was held in New York as scheduled. A proud Lin named his typewriter "Ming Kuai Typewriter (the Chinese Fast Typewriter)" and called his invention "my gift to the Chinese" ("Lin Yutang's Invention"). The major newspapers in New York City reported this invention. For the next three days, Lin opened his Manhattan home for a public exhibition of his newly invented typewriter. Hoping to commercialize his invention, Lin contacted a number of companies about manufacturing the Chinese typewriter. However, due to the civil war plaguing China at the time, no investor was willing to invest in this new invention and take the risk in marketing this product in a very unstable Chinese market. Thus, an invention that cost Lin's life's savings, about $400,000 (according to Zhaopeng Wang), never led to any financial gains for Lin and his family.

Nevertheless, news of Lin's invention did travel to China. Major newspapers in Shanghai ran reports about his invention, and, ironically, rumors were rampant about how Lin made a fortune with this invention. As unfortunate as the fate of Lin's Chinese typewriter was, this invention did find its impact in some way. First, the keyboard design was used in IBM's Chinese-English translation machine and Itek's electronic translator ("Lin Yu-Tang's Special Invention Showroom"). In addition, "MiTAC also devised the 'Simplex' Chinese character input method from the Instant Index System, which has now become one of the most common Chinese input methods in computers, widening the impact of Lin's invention" ("Lin Yu-Tang's Special Invention Showroom"). Later, this invention did find its way into the commercial market in China and became a piece of standard office equipment in larger companies and government agencies in China in the 1960s. Of course, what should be noted is that these commercial Chinese typewriters were quite different in design from Lin's Ming Kuai Typewriter.

The Chinese Typewriter

The Working Mechanism

The Chinese typewriter is fundamentally different from its Western counterpart in design and working mechanism. Other than some remote proximity in their working mechanisms, the Chinese typewriter and the Western typewriter share little in common in either appearance or the execution of the working mechanism. Figure 3 is a picture of a very typical Chinese typewriter.

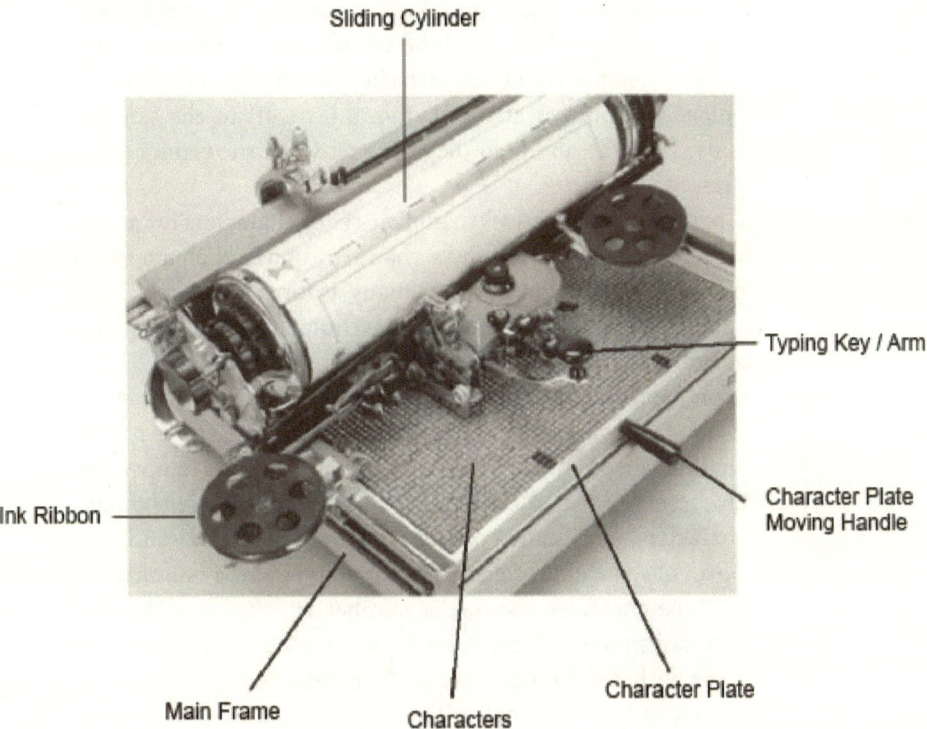

Figure 3. A Chinese Typewriter

A typical Chinese typewriter consists of the following main parts (see Figure 3):

- The main frame—This is a metal frame that holds the main parts of the typewriter together. It consists of some sliding

tracks, which allow left-right and forward-backward movements for both the cylinder and the character plate.
- A sliding cylinder—This cylinder holds the typing paper or, in most cases, the waxed stencil paper for the purpose of multiple-copy printing after typing is done. The cylinder can rotate or slide so that the typist can pinpoint the exact location for typing on the paper.
- The typing key/arm—Unlike the Western typewriter, which often has fewer than 50 typing keys, the Chinese typewriter consists of only one typing key, which has an extended arm with a knob on the top for the typist to hold and press down when typing a character. Attached to this typing arm is a holder and pointer at the bottom a little off to the side that is used to grab and hold the character when the typing arm is pressed.
- Characters—A relatively complete set of characters for a typewriter would consist of more than ten thousand characters. A complete set would include as many as forty or fifty thousand characters, and few typewriters are equipped with all the characters in the Chinese language.
- Character plates—Each typewriter uses one main character plate with a few spare ones. Each character plate is of rectangular shape and consists of several thousand empty-hole character holders. A typical plate holds two to three thousand characters, with some holding as many as four thousand.
- Boxes of spare characters—Since each typewriter has only two to three character plates, the number of characters they can hold is limited. Therefore, a typical Chinese typewriter will come with a few boxes of spare characters.
- An ink ribbon (optional)—The ink ribbon is not always found on Chinese typewriters as most such typewriters are not used to produce single-copy documents. Instead, waxed paper, which doesn't require ink, is often used for multiple-copy printing after typing.

There are certainly multiple other, minor parts on the Chinese typewriter, such as line spacers and character spacers. However, the main parts mentioned above are the standard parts of almost all Chinese typewriters, although minor parts can vary from typewriter to typewriter.

Although in theory the Chinese typewriter works in much the same way as the Western typewriter, the execution of the working mechanism is quite different. The typing itself is a rather complicated maneuver. First, the typist locates the particular character needed and moves the typing key directly over the character. Then, the typist pushes down the typing key, at which point, the pointer underneath the character plate, which is part of the typing key mechanism, pushes the character up so that the holder grabs and holds the character. The holder then swings up until it hits the paper on the cylinder to make the impression. The typist then releases the pressure on the typing key so that the key, the pointer, and the holder all return to their original positions. The typing of each character is a repetition of this multiple-step procedure.

The most complicated, and potentially most time-consuming, part of this procedure is the locating of the character needed, which requires familiarity with and memory of the arrangement of characters on the character plate. This is part of the job that requires the most training and practice. One of the most basic training components for new typists, then, is character arrangement. One may wonder if there is a fixed arrangement for all Chinese typewriters, just like Western typewriters, upon which each letter, number, or symbol is almost always located in the same positions. At first glance, the characters seem to be placed in the same way on any typewriter (see Figure 4). A closer examination, however, reveals that the arrangements can be widely different from one typewriter to another.

X	X	X	X	X	X		X	X	X	X	X	X	X	X		X	X	X	X	X	X
X	X	X	X	X	X		X	X	X	X	X	X	X	X		X	X	X	X	X	X
X	X	X	X	X	X		X	X	X	X	X	X	X	X		X	X	X	X	X	X
X	X	X	X	X	X		X	X	X	X	X	X	X	X		X	X	X	X	X	X
X	X	X	X	X	X		X	X	X	X	X	X	X	X		X	X	X	X	X	X
X	X	X	X	X	X		X	X	X	X	X	X	X	X		X	X	X	X	X	X
X	X	X	X	X	X		X	X	X	X	X	X	X	X		X	X	X	X	X	X
X	X	X	X	X	X		X	X	X	X	X	X	X	X		X	X	X	X	X	X
X	X	X	X	X	X		X	X	X	X	X	X	X	X		X	X	X	X	X	X
X	X	X	X	X	X		X	X	X	X	X	X	X	X		X	X	X	X	X	X

less commonly used   most commonly used   less commonly used

Figure 4. The Generic Appearance of Character Plates with Characters in Set

To understand the complexities of character arrangement, one must first understand the nature of Chinese characters and how they form words to make meaning. Each character has its own meaning, but it can also be combined with other characters to form different words with different meanings. For example, the character *zhong* 中 (center) can be combined with *jian* 间 (room) to mean "middle," with *guo* 国 (country) to mean "China," with *fan* 饭 (meal) to mean "lunch," and with *xue* 学 (learning) to mean "middle school." Open any Chinese dictionary, and you will find dozens of other characters that can be combined with 中 to form different words. What this means for the Chinese typewriter is that, when considering character arrangement, ideally, you want to arrange the characters in such a way that all those characters related to one another that can be combined to form meaningful words are placed in close proximity to one another. For example, when considering the placement of the character *zhong* 中, we would want all the other 30 or 40 characters that are commonly used to form word combinations with *zhong* 中 to be placed fairly close together.

Such an arrangement seems logical enough until, that is, when we consider the fact that each of the characters used with *zhong* 中 to form meaningful word combinations also has its own family of related words. The character *guo* 国, for example, can easily find its own kinship of 30 or 40 characters. Logically, these characters should be placed in close proximity to one another as well. When you consider the fact that this is true of practically all commonly used characters, with the only exception that some characters have larger kinships than others, the possibilities of character arrangement are endless. Figure 5 shows nine characters occupying a square grid. Most any two of these characters can combine to form a word. In the figure, only nine combinations are shown, which, however, should not be interpreted as the only possible combinations. In fact, there are quite a few more combinations, but for the sake of clarity, only part of the total number of combinations is shown. Imagine the number of possible combinations when you have a few hundred or even thousand characters on a single plate. Taking into consideration these endless possibilities, it's not hard to imagine the endless possibilities of character arrangement on the plate.

Also complicating the issue is the fact that different settings for typewriter use will require very different character sets due to the vastly different vocabularies used. For example, a typewriter used in an

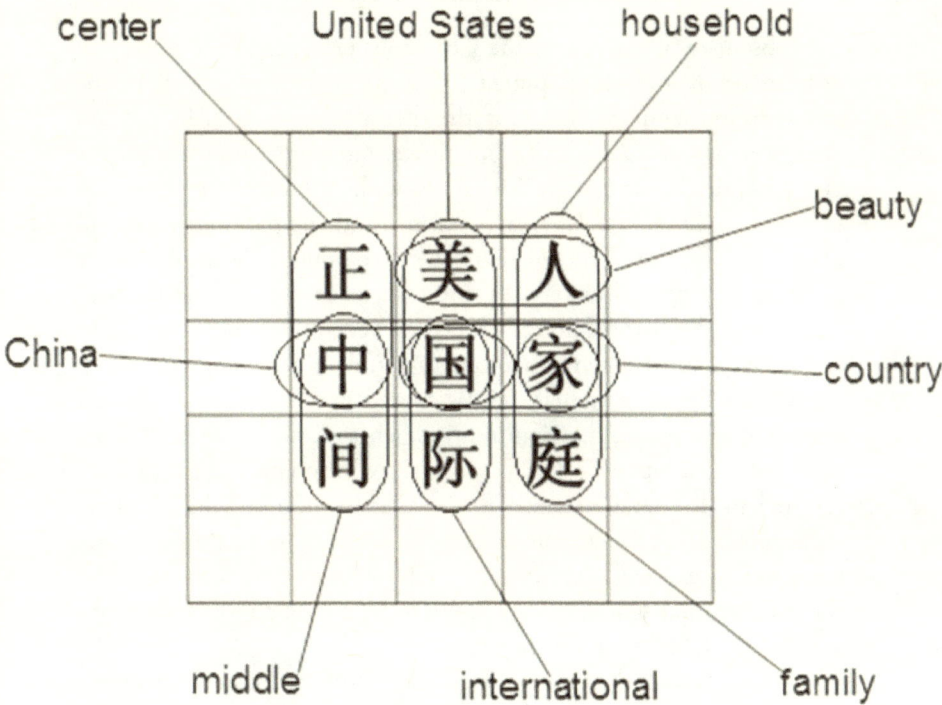

Figure 5. Sample Character Arrangement on a Chinese Typewriter

automobile manufacturing company will require a very different character set from a typewriter used in an arts and crafts export corporation. It is the typist's job, therefore, to figure out the most appropriate character set and the most logical character arrangement. Due to the endless possibilities in character arrangement as discussed earlier, each typist's arrangement is uniquely his/her own and may not make any sense to another typist.

With all these complexities, the operation of the Chinese typewriter thus requires a steep learning curve and very special training. The uniqueness of each setting for typewriter use also prevents the easy mobility of typists from one profession to another. It is no wonder then that the Chinese typewriter never became a household writing tool, although it did find its way into large office settings.

The Rhetorical Context

The obscurity of the role of the Chinese typewriter in the history of writing technology development in China renders practically futile any serious attempt at analyzing the rhetorical context of its use and later development. As pointed out above, the cumbersomeness of the typewriter itself, the complexities in its working mechanism, the steep learning curve required for its operation, and the prohibitive cost all contributed to the ill fate of this otherwise potentially revolutionary piece of writing technology. In addition, even though the news of Lin's invention of the Chinese typewriter received its due coverage in China, it quickly slipped into anonymity and oblivion, courtesy of the turmoil in China in the late 1940s caused by the civil war between the Kuomintang and the Communists, which came at the heels of an eight-year Anti-Japanese War. This almost continuous 10-year turmoil resulted in dire circumstances for investment in any field, let alone writing technology development. It was not until more than a decade later, well after the Communists seized power in 1949 and stabilized the country, that serious efforts began to be dedicated to the development of the Chinese typewriter.

One issue worthy of note, which seems to have escaped the attention of researchers in this area, is that, as insignificant a role as the Chinese typewriter may have played in the long history of writing technology development in China, its very impracticality provided, at least to some extent, a strong impetus and exigency for an important later invention/foreign transfer—the computer. Although the computer is by no means a Chinese invention, the localization of computer systems in China and China's own development of systems to accommodate the Chinese script made at least part of the computer technology in China a unique product of its own. Who could argue with utmost confidence that the ability of Chinese computers to produce Chinese characters through various input methods using the Western keyboard was not a response, at least in part, to the very inability of the Chinese typewriter to solve the same problem?

9 The Computer and the Internet

> A ragged man stands idly with a board on which is written the word 'beg.' But people just keep passing him by. "Why not add a dot com after it?" someone suggested. The man's business immediately soared after the inspiration. Okay, then, how about adding an "e" in front of the line? Soon the man became sought-after as a venture capitalist. (Lu, J., 2000)

This was a popular joke going around among the Chinese in the first year of the new millennium. Though a joke, it is rather reflective of the kind of mindset the Chinese have about the role of the computer and Internet technology in their lives. It also reveals much about how far this new technology has come along in a country that adopted it only a couple of decades ago. The fact that the development of the computer and Internet technology in China has become such a phenomenon in such a short period of time itself warrants an in-depth examination. Another reason for my focus on this development is its dual nature: this development is characteristic of most technological advances in the history of China, yet it is unique in many ways in its own right.

On the one hand, the development of the computer and Internet technology in China in the last couple of decades has exhibited some of the familiar characteristics of earlier developments: the existence of urgent social exigencies for the technology, the government's active involvement and monopoly, and the rhetorical forces that have powered this development. The social exigencies for this technology, as I will explain later in this chapter, were multi-faceted. Economically, in the late 1970s, China's economy, which had been largely stagnant in the three decades of the communist reign, called for major progress. Politically, the Communist government realized that economic improvement had become a necessity rather than a luxury for extending its hegemony. Practically, the development of this technology in the world and the need of the Chinese people to access this technology made its adoption and development in China an inevitable outcome.

A second familiar feature of this development is the government's role. As in other technological advances in China, the government's active involvement provided the major impetus for the development of this new technology. The government's motive in utilizing this technology for political purposes has produced both positive and negative impacts on the technology: a speedy development and a government monopoly of the various aspects of the development process.

A third familiar aspect is the rhetorical forces that have moved this technology forward. The participants in this development process, mainly the government and the general public, have used various rhetorical means in constructing the meaning of this new technology. The respective role playing by the participant groups and the (sometimes-subtle-sometimes-not-so-subtle) negotiation between them about the meaning of this technology have rendered the computer and Internet technology development a rhetorical process.

On the other hand, this development is also unique in its own right. For one thing, the rate of this development in China is unprecedented. Although this speedy development is true of the computer and Internet technology in the rest of the world and China's case could be viewed as a natural result of this world phenomenon, the unlikely combination of an unusually bureaucratic society in China and an unusually quick pace to catch up despite such a late start makes it a worthwhile subject of examination, to say the least.

Another unique feature that distinguishes this new development from the previous technological developments in China's history is that computer and Internet technology is a combination of foreign transfer and native development. This element of foreign transfer, which was absent in the previous technological developments that I discussed in the preceding chapters, is bound to have brought many new characteristics to this development.

COMPUTER DEVELOPMENT IN CHINA: A BRIEF HISTORY

Unlike the writing technologies discussed in the previous chapters, the computer is not a Chinese invention, although the redesign of certain aspects of the computer by the Chinese, such as the input methods, has been an indigenous innovation. Thus, computer technology in China is a combined product of foreign transfer and native development.

The Computer and the Internet

Since its introduction in China in the late 1970s, the computer has been developing at a surprisingly rapid rate, to say the least, especially in recent years. Of course, like any other technology, the computer went through several rather distinct stages of development: experimentation from the late 1970s to the early to mid-1980s, acceleration from the mid-1980s to the mid-1990s, and exponential growth from the mid-1990s to the present. A knowledge of this development path will help us understand what elements have effected this development and rendered computer technology what it is today in China.

The Period of Experimentation (From the Late 1970s to the Mid-1980s)

China's exposure to computer technology did not begin until the late 1970s. The first 500 microcomputers were imported into China in 1979 (Liu, 1987, p. 190). Although the Chinese leaders were eager to boost the new technology, they were cautious at first since they had no idea about the potential consequences of this new technology. Thus the first several years were mainly devoted to experimentation in developing computer technology. Even with this careful control by the government, this period saw some good growth in sales. Figure 6 shows a steady growth in computer sales from 1982 to 1986. Although

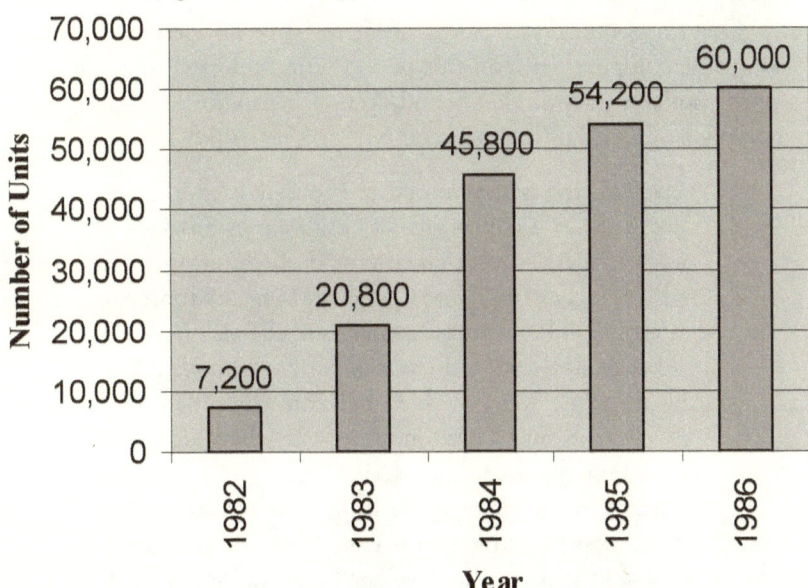

Figure 6. Increase in Computer Sales in China (1982–1986). (Zhang & Wang, 1995, p. 13).

the sales revenue shows a slight drop from 1985 to 1986, it was due to the decline in prices as a result of increased supply and competition (Zhang & Wang, 1995, p. 12).

This was a period of controlled development. On the one hand, the government was wary of the unknowns about this technology; on the other hand, it was eager to understand it, to take advantage of it, and to realize its great potential. This eagerness was evident in the government's technology policies and orientation during the period. As stated by Fengqiao Liu (1987), Director of the China Software Technology Development Center and Deputy-Director of the New Technology Department of the State Science and Technology Committee, "since 1980 the SSTCC (the State Science and Technology Committee of China) has promoted research and development (R&D) into Chinese information processing technology," and much of this R&D effort was put into developing computer technology, including "the standardization of Chinese characters and Chinese character input, processing, display and output technologies" (p. 191). Therefore, by the end of this period, China's computer market, though still relatively small, was of considerable size when compared with that in 1982. Although Liu (1987) puts the total number of computers in China by 1987 at about 137,000, by Zhang and Wang's calculation, obviously the actual figure should be well above that.

One important development needs to be mentioned here. Nearly all of the computers sold in this period were imported. However, toward the later part of this period, China began to develop its own computers. Li (1996) has this account:

> In 1985, the State Computer Industrial Administration selected a contingent of backbone technicians to form a scientific research task force which was supplied with adequate funds and a good working environment. After feasibility studies, members of the task force decided to develop products compatible with IBM and PC systems. Some of them built an experiment base in Hong Kong. Three months later, they succeeded in debugging gate-array design series products. Another team in Japan made important breakthroughs in developing Chinese display software. By June 1985, they had successfully developed a domestically-made PC computer, Great Wall 0520Ch, which was the first

with a Chinese language display function using the method of character generation, and with complete information processing capacity in Chinese. (p. 11)

In fact, China's first efforts at developing its own computer began as early as 1974 (Li, 1996, p. 11). However, these efforts ended in failure due to various problems. The successful design of the Great Wall 0520CH computer marked the real beginning of China's PC industry and its native development of this technology. Soon after that, this model began to be produced on a large scale, and it "gained a large share of the domestic market" (Li, p. 11).

The Period of Acceleration (From the Mid-1980s to the Mid-1990s)

This second period saw an accelerated growth in computer sales and development in China, especially in the 1990s. During this period, with the establishment of several domestic computer companies and the development of various domestic models, China's PC industry began to grow out of its infancy. Figure 7 illustrates the rather tremendous growth, especially in the early to mid-1990s. For example, after moderate increases in sales each year from 1987 to 1990, the number of PC units sold in 1991 increased by more than 100 percent compared with that in 1990. The increase over the next few years, though not nearly as much, maintained its strong trend.

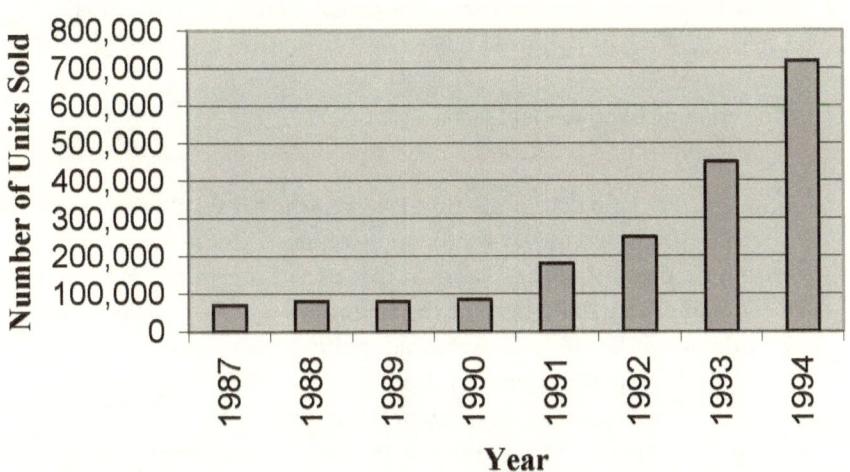

Figure 7. Accelerated Growth in Computer Sales (1987–1994). (Zhang & Wang, 1995, p. 13).

A more interesting development is in the evolving patterns of China's import and export of computer products during this period. As is shown in Figure 8, though both import and export numbers generally maintained an increasing trend over the years, export, which had been less than import each year until 1992, surpassed import for the first time in 1993. What this means is that by the early to mid-1990s China's PC industry was rather well established and China was beginning to free itself from its dependence on foreign technology. If the 1993–1994 growth rate for export (79.5%), which is far greater than that for import (34%), is any indication, the domestic industry was obviously going strong.

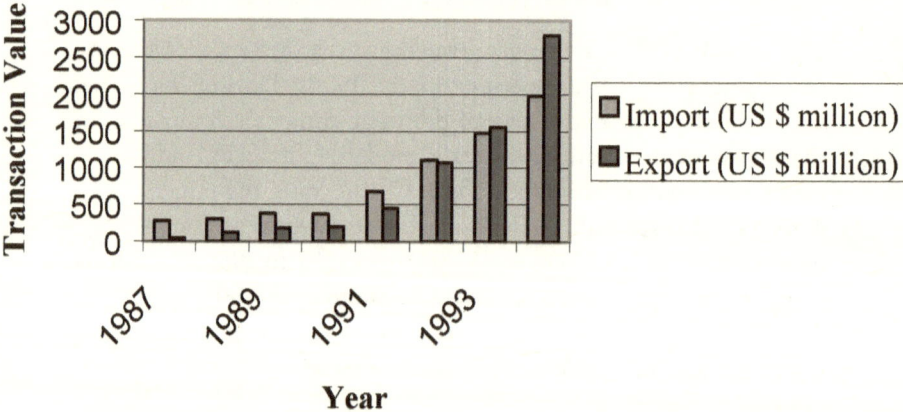

Figure 8. China's Import and Export of Computer Products (1987–1994). (Zhang & Wang, 1995, p. 14).

The Period of Exponential Growth (From the Mid-1990s to the Present)

Although the period from the mid-1980s to the early 1990s saw considerable growth in the development of computer technology, it must be pointed out that during most of this period except the later part the computer remained beyond the financial reach of most families. Even though 718,000 units were sold in 1994, this figure still seems minimal when we consider the 1.2-billion population of China. It was not until the mid-1990s that computers became accessible to the common people due to a variety of reasons, among which were increased income and living standards and substantial price decline as a result of lower cost and increased competition among computer manufacturers. Computers began to enter the ordinary household.

In 1995, the number of PC computers sold jumped to 1.2 million as compared with the 718,000 of the year before (Ren, 1997), an increase of 67 percent. This figure rose to 1.8 million in 1996 (Yan, 1997) and 3.5 million in 1997 (*PRC Yearbook '98/99*, p. 246). Though no figures are available to the researcher for 1998 and 1999, we have reason to believe that this exponential growth has kept up in China as in the first quarter of 2000 alone, PC sales reached 4.1 million ("Asian PC," 2000; Hou, 2000a). Even in terms of sales revenue, the increase has been substantial, though prices have generally declined. China recorded ¥80 billion RMB in computer sales in 1996 (Yan, 1997) and ¥200 billion in 1999 ("Domestic brands," 2000).

Not only have sales skyrocketed, but sales of higher-quality computers are also increasing. While prior to the mid-1990s the computers ordinary people owned were more likely UNIX or DOS systems, most people's computers today run on multimedia platforms. Statistics show that by 2000 the Chinese owned at least 10 million multimedia computers (Zhu, 2000). By 2007, the estimated total number of personal computers in China reached 50 million ("Vista Changes"). This number, however, is a rather conservative estimate since the most recent comprehensive survey of Internet development in China by the Computer Network Information Center of the Chinese Academy of Sciences finds the total number of computers in China connected to the Internet alone to be 59.4 million ("Survey of Internet Development, No. 19," 2007). Therefore, the total number of computers in China is probably well over 60 million.

Internet Development in China: A Brief History

The rapid growth in computer development is matched by a corresponding development of Internet use in China. Though the history of Internet use in China is much shorter than that of computer use, its development rate has far surpassed that for the computer.

Rapid Growth in Chinese Internet Users

The Internet did not emerge on the Chinese scene until the late 1980s. The first use of the Internet in China was in 1987 (Xiao & Yang, 2000). However, the Internet to most Chinese at that time was still an alien concept. The ordinary Chinese did not start using the Internet until the early to mid 1990s. The number of Internet users in 1993 was a mere 1,700 (Qin, 1997). However, despite this late start in Internet development and use, China has been mounting a ferocious effort to catch up with the rest of the world, and the growth rate and growth potential have been tremendous. In 1996, the number of Internet users in China was 120,000 (Qin, 1997). This number increased by five times to 620,000 the next year. It jumped to two million in 1998. After that, the number doubled every six months in the first few years, resulting in 22.5 million total Internet users by January 2001 and a staggering 137 million by January 2007. The number of computers connected to the Internet grew from 3.5 million in 1999 to 59.4 million in 2007 (see Figure 9), and the number of websites based in China increased from 15,153 in 1999 ("Quadrupled," 2000) to 843,000 in 2007 ("Survey, No. 19," 2007).

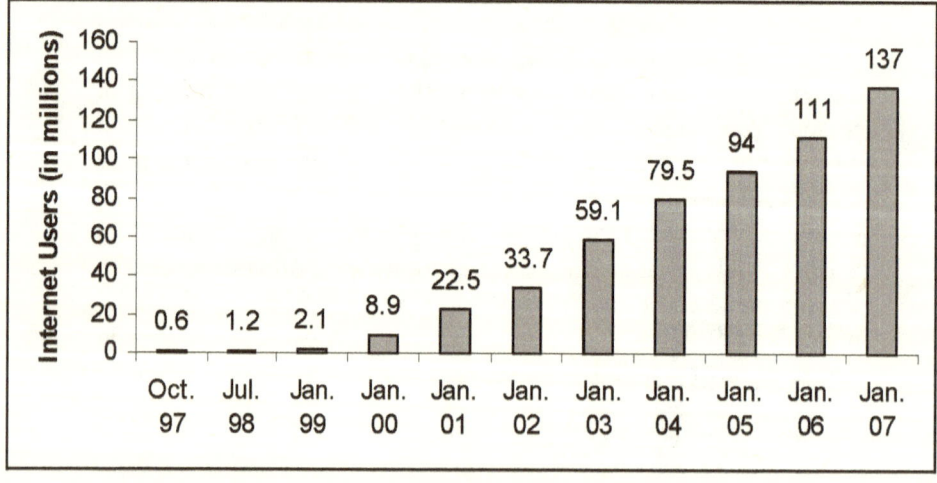

Figure 9. Increase in the number of Internet users in China (1997–2007). (CNNIC)

An interesting fact about Internet development in China is that the growth rate has constantly surpassed people's predictions. For example, in 1997, it was predicted that by the year 2000 the number of Internet users in China would grow to 1 million (Qin, 1997). What actually happened went far beyond that prediction. The number jumped to 4 million in June 1999 and 8.9 million by the end of 1999 (Pomfret, 2000). Then in his article in February 2000, Pomfret predicted that the number of Internet users in China by the year 2001 would reach 12 million. Little did he know that the number would reach almost 17 million in a mere four months after he made the prediction. It should be no surprise, then, that this figure had reached 137 million by January 2007.

Major Developments in China's Internet Use

Apart from the rapid increase in the number of Internet users, another interesting aspect worth noting is the major developments in the early history of Internet use in China, which reveals how and to what extent the Internet has been used at various points in the first few years. Though, as I mentioned above, the first Internet use in China was as early as 1987, the Internet did not become visible until after the mid-1990s. Xiao and Yang's (2000) "The Firsts of the Internet in China" provides a brief account of the major ground-breaking events in Internet development in China. Though not many details are provided for these firsts in Xiao and Yang's article, a description of these early events is meaningful here since they provide a window to the various stages of Internet development in China and give us a sense of the level of penetration into various fields in the lives of the Chinese.

- The First Chinese to Use the Internet (9/20/1987)
- The First Network to Connect to the Internet (4/1994)
- The First Chinese Media to Go Online (10/20/1995)
- The First Internet Regulations (1/23/1996)
- The First Distance-Education Class (1996)
- The First E-Café/E-Bar (11/1996)
- The First Online Magazine (1/1997)
- The First TV Series to Go Online (1998)
- The First Chinese Search Engine (2/15/1998)
- The First Large-Scale E-Commerce (6/1998)
- The First Hacker (6/16/1998)
- The First Domain Registration in Chinese (1998)
- The First Online College (11/6/1998)
- The First Government Web Site (12/16/1998)
- The First Pornography Recognition Software (4/8/1999)
- The First Internet-Related Intellectual Property Case (4/28/1999)
- The First Internet TV Station (6/1/1999)

These first events are significant in that they provide an insight into the way the Internet penetrated into the various fields in China in the first few years. For a discussion of the details of these events and their significance, see Appendix A.

Perhaps, the most significant aspect about computer and Internet development in China is the great leap in writing technology and medium they represented. For a culture accustomed to manual writing with pen and paper for almost four thousand years, the advent of the computer and the Internet was nothing short of a revolution in writing technologies and media and facilitated the transition from manual to mechanical writing, a transition that, in the Western cultures, had been completed more than a century before, when the Western typewriter came into existence. Even though a substantial part of the Chinese population is still relying on the old pen and paper medium for their written communication, the computer has already become the dominating medium of written communication for the younger generations of today.

The Rhetorical Context

While it is interesting to speculate on the future of computer and Internet technology in China, the focus of this study, however, is more on what prompted this development than on where it is going, though the latter is inevitably a part of the study. Therefore, this section will focus on how the six elements of my model of rhetorical analysis, namely, exigency, ideology, participants, knowledge construction, knowledge control, and communication medium, have been playing their role in this development process.

Exigency—Economic, Political, and Practical

Though computer and Internet development in China has been in part a result of development trends worldwide, several uniquely Chinese factors have contributed to creating the exigency for such a development. These factors include economic necessity, political expediency, and practical viability.

Economic Necessity. As mentioned earlier in this chapter, the Communist rule until 1976 had been one of extremely impoverished economic growth due to various disastrous political movements that culminated in the Cultural Revolution. With the demise of the Mao regime in 1976, and after a brief transition period of a couple of years, Deng Xiaoping came to power, which marked the beginning of the post-Mao era of economic reconstruction. A milestone in China's policy making took place in 1978, when the Third Plenary Session of the Eleventh Central Committee of the Communist Party of China was held, which not only negated the Cultural Revolution and much of the Maoist regime, but, more importantly, established the open-door policy for economic development. Also established was the country's long-term goal of modernization in four key areas: agriculture, industry, military, and science and technology. Of these "Four Modernizations," which became the fundamental national policy for the next two decades, the modernization of science and technology was placed in the priority position because the Chinese leaders recognized that science and technology were the key to economic, and in fact all other, development.

This open-door policy and emphasis on economic development triggered China's thirst for advanced technologies, much of which were available only in other countries. Foreign technologies began to pour in. Under such historical circumstances, the introduction of the first computers into China in 1979 was therefore more than a mere coincidence. With computers being the symbol and representation of most advanced technologies, the introduction of computer technology marked the real beginning of a serious effort to rebuild the much damaged economy. What happened in the next two decades was nothing short of an economic miracle, which accounts for the computer and Internet technology in China today.

The main reason for such a change is China's transition to a market economy as a result of the economic reform that started in 1979. According to Zhang & Wang (1995), this reform is characterized by the following four features: (1) gradualism and experimentation, (2) successful partial reforms within certain sectors, (3) decentralization of decision-making power to enterprises, local governments, and individuals, and (4) self-reinforcement (p. 5). The caution and experimentation at the beginning stage of this reform accounted for the relatively slow development of computer technology in the early 1980s. Successful partial reforms within certain sectors ensured a balance between development and control. The decentralization of power provided a major impetus to economic development in the late 1980s and early 1990s. The self-reinforcing effect resulted in the spread of reform from one area to another. This economic reform and transition to a market economy also featured enterprise reform (privatization of ownership), financial and fiscal reform, further reforms on trade and investment regimes, and price and social sector reforms (p. 6). This economic reform resulted in GNP growths of 3–4 percent in the 1980s (p. 5) to close to 10 percent in mid-1990s, as compared with close to zero percent in the period between 1949 and 1979. Right now, although state-controlled economy is still existent, the major elements of a market economy are in place.

Therefore, if the initial development of the computer technology in China was more of a government initiative, by the 1990s, especially the mid to late 1990s, it had become at least as much a logical result of economic development as a strategic orientation. Much of China's economy had evolved from taking advantage of computer technology to depending on it, which is more true than ever today. The main difference between today and the late 1970s and the early 1980s in computer development is that computer technology has become an economic necessity instead of a luxury aid. As Iritani (2000) points out,

> The Internet could provide a powerful boost to the nation's economic reforms by streamlining the nation's banking and manufacturing systems. China's manufacturers, for example, must rapidly upgrade their technology or lose out to competitors as the world's largest automotive, apparel and electronics firms move their supply chains online.

Indeed, without embracing the Internet, China stands little hope of catching up with the industrial world. Thus the government is modernizing its telecommunications and computer infrastructure, while trying to lure back would-be tech entrepreneurs who studied abroad.

Therefore, for many businesses, adopting computer technology is now more a survival strategy. The same is becoming more and more true for China's whole economy.

Political Expediency

If economic development had been the only contributing factor, the argument that it alone could have led to the exigency for computer development would be hardly convincing. Chinese leaders have a history of pursuing political gains even at the cost of economic development, and the new leaders since Deng, although not quite the same as Mao, have yet to prove how much different they are from that political tradition. Political expediency has been a major player in creating the exigency, though it is one that is only tacitly and never openly acknowledged. In some ways, it has been the drive behind the economic reform, which in turn led to the development in computer technology.

By political expediency, I mean the necessity and the means to ensure and consolidate one's political position and power. There are two aspects to this political expediency in the case of China in the past two decades: domestic and international. Domestic political expediency basically refers to the need to confirm and strengthen the ruling position of the Communist Party while international political expediency refers to China's need to regain its role as a major player on the international political stage. Both aspects have been vitally important to the Chinese leaders in the past two decades.

During the Deng and later Jiang (Zeming) eras, domestic political expediency became an issue, unlike in the Mao era. Mao's rule of China found its legitimacy in Mao's being the founder of the Communist government. With China's doors closed to the outside world, Maoism dominated the political ideology. People were taught to harbor ultimate respect for Mao as their leader, and they never questioned the legitimacy of Mao's leadership. Lack of exposure to the concept of Western democracy on the part of the Chinese people led to their blind love for their "beloved leader" even when he was making mistakes. As a result, Mao achieved a status equal, and in some ways even superior, to that of an emperor. During his reign from 1949 to his death in 1976, Mao's rule was never endangered, although he did experience some minor threats from a few of his close comrades who had fought with him and helped him establish the Communist government in the 1940s.

When Deng came to power in the wake of the Cultural Revolution, the Chinese people were beginning to see the real nature of this, perhaps the greatest, human tragedy in the history of China. Exposure to the ideas of Western democracy as a result of the open-door policy prompted them to question certain aspects of the Communist rule, one of which was its economic policies. Deng and his comrades could no longer legitimize their rule simply with the dogmatic ideological brainwashing that had carried Mao through his tenure. As the people increasingly opened their minds to Western ideas, they more and more strongly demanded a government that would bring them a better, improved life. Obviously, the Communist Party leaders realized that a better economy not only would strengthen their power but was practically the only strong means to legitimize their rule. Such a realization resulted in a series of policies conducive to economic reform, which in turn led to a great impetus for technological development.

Internationally, political expediency also called for an economic reform. During Mao's rule, the closed doors and his disastrous economic policies rendered China a weak giant on the international scene. When Deng and other Communist leaders opened China's doors to the outside world, they were not merely interested in advanced foreign technologies: they were eager to regain China's role as an important player on the stage of international politics. To regain such a role required a strong military. A prerequisite to a strong military was a strong economy. The disintegration of the Soviet Union and its economy in the 1980s and its ensuing loss of prominence in international politics served as a sobering reminder for the Chinese leaders that a strong economy was the key if they wanted to avoid the Soviet path.

Both domestic political expediency and international political expediency prompted the Chinese leaders to put an emphasis on the development of computer technology. Fengqiao Liu's (1987) description confirms such a general orientation:

> China is a developing country. Its government has already recognized that 'the third industrial revolution' constitutes both a challenge and a perfect opportunity for the modernization of the country. China can make immediate use of the newest scientific and technical achievements and 'leap-frog' some conventional stages of development, such that the process of modernization and the improvement of the material and spiritual well-being of its people can be accelerated.
>
> In order to meet the challenge of this revolution China must take as its starting point the existing national conditions. . . . Traditional industries must therefore continue to play a major role in the Chinese economy for some time to come, in order to meet the

> basic daily needs of the people. However, the Chinese Government is giving priority to the use of microelectronics and computer technology in order to speed up the technical transformation of traditional industries. (p. 189)

Given Liu's official position (Director of the China Software Technology Development Center and Deputy-Director of the New Technology Department of the State Science and Technology Committee), his words are more than personal opinion but are instead representative of governmental policies. Such official efforts to boost the computer industry have been consistent in the last two decades.

Practical Viability. A third, though relatively less deciding, contributing factor for the exigency for computer development was its practical viability. If, in the late 1970s and the early 1980s, computer technology were limited only to a few select sectors and the common people still found it a rather alien concept, by the late 1980s and the early 1990s, people began to recognize the practical functions of the computer to change their lives. Coupled with a strong desire to improve their lives due to economic reform, such a recognition began to turn the computer from a high-brow technology into a popular multi-function practical device in everyday life. It was then that the computer began to enter common people's lives. As Li (1996) describes it,

> Personal computers have gradually entered the life of the Chinese people. Reporters and writers were the first to introduce computers into their families and offices to replace pens. Later, the public gradually found that computers offered extra ability in education and housekeeping along with other attractive superior functions such as entertainment, telecommunications and information networks. This has led to multi-functional computers entering a growing number of ordinary families, functioning as CD and VCD, telephone and fax, TV set, educational tool, man-to-machine talking, and network operation.... (p. 10)

In addition to the practical multi-functionality of the computer to improve the quality of life, another factor that contributed to the practical viability was the nature of the Chinese language. Before the introduction of the computer, written communications in Chinese were largely limited to the handwriting mode only. The Chinese typewriter, due to the impracticality of its design as we discussed in the last chapter, never became a popular tool in people's everyday life. With the emergence of the computer, what used to be a luxury option and required professional service suddenly became possible with a simple computer and a printer. It was not surprising then that the computer became popular quickly in China in the 1990s.

If economic necessity and political expediency were the two exigencies that initiated and provided the main driving force for computer and Internet development in China, it was the exigency of practical viability that popularized this new technology. Granted that lack of the first two exigencies would have rendered computer technology an impossibility in the social context of China, an absence of the third exigency might have reduced the computer and Internet development to something less than the phenomenon that it is today.

Competing Ideologies—Traditional, Political, and Western

The influence of ideology on computer and Internet development in China is probably one of the most difficult topics in this study. Important as it is, it is almost impossible to describe the influence of ideology in articulate terms. What follows then is at best only an attempt to understand what ideological orientation has driven the participants to look at the computer and Internet technology the way they do. Given the complex nature of the Chinese culture at present, it seems that three kinds of ideologies have affected people's perspectives on the new technology: traditional ideologies, such as Confucianism, Taoism, and Buddhism; political ideologies, such as Maoism and Marxism; and Western ideologies.

Traditional Chinese Ideologies—Confucianism, Taoism, and Buddhism. Traditional ideologies are deep rooted in the collective cultural thinking of the Chinese society. The values they uphold and the rhetorical orientations they promote are reflected in all facets of the everyday life of the Chinese people. The depth of their influence is hard to fathom, yet their omnipresence demands due attention in this study. It is no exaggeration to say that everybody is affected by these ideologies in one way or another, although different participant groups use the ideologies to serve their respective purposes.

As discussed in Chapter 1, three of the most outstanding features of Confucian rhetoric are communalism, historicism, and dialogism. These three features are no doubt at play in shaping people's perspectives on the computer and Internet technology. As the central tenet of Confucianism, communalism promotes a rationalized social order to be realized through the Confucian concepts of *jen* (humanity), *yi* (justice), and *li* (etiquette). Portraying the computer and Internet technology development as a social need, the government has appealed to the general public to look at this technology as a tool for self-cultivation and social improvement, an important means to achieve a rationalized social order. Historicism, on the other hand, posits that what this technology means depends on a contextualized communal understanding to be reached through dialogism. One image the government constantly attempts to project in various media is its efforts to establish a communal understanding of this technology, by using the various perspectives on the meaning and potential benefits of the technology to different people: scientists, technicians, ordinary users, etc. This image of seemingly dialogic communication is designed to invoke a positive attitude toward this new technology, which is not unappealing to the general public. For the ordinary users, computer and Internet technology is indeed an effective means of self-cultivation in that it leads to greater knowledge, better jobs and life, and more respect.

The Taoist principles, though in many ways radically different from Confucianism, also seem to have a positive impact on people's attitudes toward the new technology. The unorthodox, relativist worldview of Taoism leads to the realization that everything happens for a reason. Therefore, the very occurrence of technological innovation justifies its existence. At the same time, the Taoist view of knowledge and truth as products of perception may have led to people's varying views about the computer and Internet technology and their justification of such views. In addition, the Taoist emphasis on individual initiative may also have been one of the reasons behind people's active participation in this development process, and it is certainly much used by the government as a persuasive tool. It seems both Confucianism and Taoism have played in the favor of the new technology, from both the government's and the general public's perspectives.

The effect of Buddhism on the development of the computer and Internet technology may be more ambiguous, partly due to the fact that "Buddhist rhetoric is a variously divisible school, having great potential to contain contradictions" (H. Wang, 1993, p. 67). On the one hand, the Buddhist paradoxical view that things neither exist nor non-exist as things may result in a rather passive attitude toward everything. On the other hand, its principle of equality among all people may have fostered active participation by the masses in the development of the new technology.

Limitations of space and research resources prevent me from going much further in exploring the influences of these traditional ideologies, which is much needed for an in-depth understanding of the development process of computer and Internet technology. One thing is clear, however: the multitude of ideological thoughts and their rhetorical implications have, on the whole, produced attitudes conducive to the development of the new technology. As H. Wang (1993) puts it:

> Rich traditions like those in China have great potential. The paradigmatic characteristic of Chinese traditions allows, as we see in changes of interpretation even within one school of thinking, significant revisions; its cyclical feature accommodates revising efforts with many choices and alternatives to combine different elements of thinking. These traditions, on the other hand, have also prepared an understanding and accommodating audience for discursive activities. Hardly any way of thinking/reasoning, probably, will come to the Chinese audience as a total surprise or entirely incomprehensible, including the paradoxical validation of contradictions. (p. 68)

These varying schools of thought, indeed, prepared the Chinese for the new development.

Political Ideologies. Political ideologies, though not as deep-rooted as the traditional ideologies mentioned above, have been playing an important part in shaping people's attitudes toward the new technology. Marxism, the most dominant political ideology in China since 1949, has been used by the Communist leadership, in selective partiality rather than its entirety, to serve its ulterior purposes. The Chinese government has used mainly two aspects of Marxism: its concepts of surplus value and economic exploitation and its principles on eliminating social stratification. In the first 30 years or so, the Marxist concept of capitalist exploitation was the official propaganda against financial inequalities among the people. In the last two decades, however, this concept seems to have been used in a reverse direction. Although the government never denounced Marxism, both the government propaganda and the popular perspective at least in the early days of China's open-door economic policy (i.e., the 1990s, seemed to be that anyone was entitled to the opportunity to make surplus value and become wealthy. Financial inequality was no longer a problem, but something encouraged by the government. Along the same line, although the elimination of social stratification is never officially promoted, the creation of financial inequalities effected the distinction of social classes. It was not until recent years that the Chinese government realized the potential negative consequences of such social and economic inequality and sought to contain and reverse the inequality trend. Ironically, however, Marxism is still hailed as the ideological tool to "educate the people."

Maoism inherited mainly the two aspects of Marxism mentioned above, with the addition of some principles specific to the Chinese context. One notable addition is the principle of self-reliance, which is still followed today, though in a much modified way. The concretization of this principle in computer and Internet development is China's emphasis on the development of China's own software and hardware technologies. This is reflected in the generous support by the government for its own IT industry.

The revised political ideologies, especially those by Deng Xiaoping, are oriented toward more open policies on economic development, most notably reflected in its adoption of the capitalist system of market economy, though the government has never acknowledged its capitalist nature. As mentioned earlier in this chapter, the loosening of ideological control as a necessary justification for the new Communist economic policies made technological development possible. In the process of opening up, however, the Communist government has never given up Marxism and Maoism, which are still used as ideological measures in its desperate attempt to contain the "negative" consequences of this new development.

Western Ideologies. While the government is trying to cling to the generations-old political ideologies, the masses are turning to Western ideologies, especially that of democracy, which is made accessible in a large part by the computer and Internet technology. On the one hand, the government is resolved "to use the Internet to support economic growth, while making provisions for maintaining political stability" (Green, 2000). On the other hand, "in a multitude of minor ways, the Internet will improve that amorphous of things, the country's democratic spirit" (Green). Among the quickest to accept such new ideologies are members of the young generation:

> As in most cultures, youths are the vanguard of this societal shift. But here, China's one-child population policy, an injection of free-enterprise ethos, plus a growing disillusionment with communism, all contribute to the creation of a generation of individualistic, pampered "little emperors." They have the means to go online to explore new values and the desire to join a "pop planet" cultural movement. (Platt, 2000)

Despite government efforts to control certain information, "the talk on the Web is some of the freest in China," and "the need for political reform is a common topic in Chinese chat rooms" (Pomfret, 2000).

This exposure to Western ideologies is not likely to effect any ideological revolutions, at least not in the near future, although it is definitely producing evolutionary changes in the way people look at things, including the Communist political ideologies. Therefore, the influx of Western ideologies does not pose a serious threat to the Communist regime at present. As Green (2000) explains,

> The reason for this is simple: the government is not all that bad. Despite its dubious human rights record, it has delivered the economic goods and maintained public order. This matters a great deal since it affects the majority of lives. The government is not perfect, but there is no groundswell of opposition, however much most people wonder what on earth Marxism has to do with their daily lives.

The long-term effect of this exposure to Western ideologies, however, is more difficult to project.

Participants—A Polarized Spectrum

In a democratic society, the spectrum of participants often consists of the following three groups: the political elite, the technical elite, and the ordinary users/the masses/the general public. These groups play different roles and function in different ways in carrying forward the development process. The role of the political elite is one of moderation, creating and implementing policies to ensure that technology development stays on a certain course. The technical elite functions to map out the theoretical aspects of the technology and to turn the technology from imagination, first, into a design possibility and, then, into a reality. The general public plays a central role among all these participant groups by defining the nature, the uses, and the purposes of that technology and by creating the exigency for the technology. It therefore shapes the development path in significant ways. Though for different technologies the roles of these participant groups will vary from context to context, anomalies are uncommon, and these groups function more or less in accordance with their capacities.

In a totalitarian society like China, however, it works differently. Participants are perhaps the most intriguing aspect in the development of computer and Internet technology in China in that it differs to a considerable extent in the way participants function in this Chinese context. There are two unique features about the China case, both of which have exerted a profound impact on the course of the development.

The first uniqueness is that there are only two participant groups—the government and the general public—instead of the usual three theorized in Western cultures. Where did the other two—the political elite and the technical elite—go? They did not really disappear but have acquired different capacities. In a democratic society, the government is part of the political elite. In the Chinese context, however, the government becomes the dominating group, with the political elite as its subordinate subgroup. Ideally, within such a context, this political elite group should function as both enforcer and mediator. In other words, it should enforce government policies and, at the same time, mediate between the government's and the general public's perspectives on the nature and the uses of the technology.

In reality, however, instead of mediating between the government and the other participant groups, the political elite becomes part of the bureaucracy. Rather than interpreting as well as implementing government policies, in China, the duties of this group are reduced to implementation alone, the implementation of government's policies and its definition of technology. It becomes a mere tool of the government. When it loses the power to interpret the policy (and thus to influence the development process), it loses its distinct identity and meshes into the government group.

By the same token, the technical elite loses its identity as a distinct participant group because it, too, is deprived of the power to define the technology. In a democratic society, the technical elite is the real bridge over the gap between the political elite and the general public since they are in a unique position to influence the decisions of the political elite and help shape and reflect the general public's perspective. In the Chinese context, however, while the technical elite may still be able to affect the general public's perspective on technology, it is not in a position to influence the decisions of the government. In this sense, the technical elite becomes more a part of the general public than a part of the government group.

Another unique feature about the participants in computer and Internet technology development in China is the shift of the central role from the general public to the government. While in a democratic society, it is the users (i.e., the general public) that determine the development path of a piece of technology, in the Chinese context, this power to define technology is taken away by the government. Unlike the government in a democratic society, whose role is largely one of moderation, the Chinese government assumes ultimate control to define the nature, the purposes, the uses, and the development path of the computer technology, and there is no sign that it is willing to relinquish this control anytime soon. Having said this, I must point out, though, that all participant groups, whatever their role, still contribute to defining technology. Let's take a look at how the government and the general public have contributed in different ways in their respective capacities to the development of computer and Internet technology in China.

The Government—The Dominating Agent. To understand the role the Chinese government plays in developing computer and Internet technology requires one basic understanding (i.e., the ultimate priority for the Chinese government in this development has always been to improve the economy while keeping the political regime, and therefore its rule, intact). Consolidating the Communist rule is always the number-one goal and number-one priority. As was mentioned earlier in this chapter, the Chinese government understands that economic development is becoming a necessary condition for achieving such a goal. However, they also understand that with economic development often comes ideological progress, which is the source of threat to the Communist rule. For Chinese leaders, the perfect result would be a combination of great economic progress and the same old communist ideology. Therefore, a guiding principle for the Chinese government in developing computer and Internet technology seems to have been to promote "the positive aspects" of the technology as much as possible, as long as it does not constitute any real threat to the Communist regime, while fending off "the undesired elements" that came with the capitalist technology. Such an orientation has been consistently evident in the series of policies and strategic moves the Chinese government has made in this development.

This balanced orientation by the Chinese government between technological development and ideological control has resulted in two seemingly contradictory categories of moves by the government: projects and policies to promote development on the one hand and laws and regulations to control ideology on the other. The strange combination of the two have produced something close to a miracle in technological development.

The government's role in computer and Internet development in China has been pivotal, to say the least. A review of this development in the last three decades reveals that except for the experimentation period from the late 1970s to the mid-1980s the Chinese government has been making a consistent effort to develop this new technology. Since the real, substantial advances in computer and Internet technology did not occur until the 1990s, my analysis of the government initiatives will focus on this last decade or so. Nevertheless, a brief summary of the relevant policies in the 1980s would help us better contextualize the policies in the 1990s. A complete account of all the policies and major decisions made by the Chinese government concerning computer and Internet development is obviously impossible. My analysis will, therefore, sample some important developments in policy making to provide an illustrative and representative rather than comprehensive picture.

As mentioned earlier in this chapter, China initiated its open-door policy in the late 1970s. This policy spearheaded an economic reform that started in the early 1980s. A significant feature of this reform has been a shift of China's economic system from a planned economy to a market economy. China's trade system, for example, "has moved from one in which all trade was planned and carried out through a handful of foreign trade corporations (FTCs) to one in which the role of planning is much diminished" (Zhang & Wang, 1995, p. 6). In December 1982, in introducing China's Sixth Five-Year Plan (1981–1985), the then premier Zhao Ziyang declared that "in the future, as with other sectors of the economy, science and technology would be increasingly governed by economic rather than administrative measures" (Wang, 1993, p. 109). A major accomplishment in the economic reform in the 1980s was the gradual monetary transformation of the economy, which provided a necessary condition for the transition to a market economy (Wen, 2000). It was under such a context that computer development accomplished its move from experimentation to moderate growth in the 1980s.

In the early 1990s, the government began to exert real efforts specific to the development of computer technology. For example, two important policies were made in this period: *State Council: Regulations on Protecting Computer Software* in 1991 and *Detailed Rules on the Registration of the Copyright for Computer Software* in 1992 (Zhang & Wang, 1995, p. 159). These two policies were significant in that they represented the government's serious effort to legalize and protect the young industry.

The same orientation was obvious in various talks given by high-level government officials on various occasions. In 1995, for example, in his speech at the National Electronics Industry Working Conference, "Hu Qili, minister of the electronics industry, called on the entire sector to actively adjust its industrial structure and strive to construct a modern electronic information processing industry in China" (Li, 1995, p. 12). Though this was not directed at the computer industry alone, it was an important orientation policy-wise given the increasingly important status of the computer sector in the electronics industry.

The Technology Plan for 1996–2000 unveiled by the Chinese government in 1995 mapped out the government's plan to use technology to boost its economy. Though this plan had agricultural growth as its top priority, its main purpose was to "promote basic scientific research into key scientific and technological issues connected with economic and social development" ("Technology plan," 1995, p. 5). Listed in the number-one position in the major industries enlisted to help achieve such a goal was information.

This consistent effort to promote computer and Internet technology continued into and even intensified in the second half of the decade. One indicator of this intensified effort is found in the news reports published in the *People's Daily*, the official newspaper of the Chinese government. One news report published on February 10, 1997, summarized the government's position about development of the software industry. According to the report, the short-term strategy was "to engage in some and to avoid some others" ("Short-term goals," 1997). What this means is that in developing its own software industry China will learn the lessons of other countries (i.e., to avoid those already mature technologies and concentrate on those projects on which China has an advantage, such as Chinese information processing technologies and applied software technologies), or is at a compatible and thus a competitive level with other countries, such as with multimedia software, educational software, financial software, and all software related to the Chinese culture.

This same report also maps out the short-term goals of China's software industry:

1. Total sales for the software and information technology industry to reach ¥52 billion RMB
2. Total number of software products to surpass 50,000 kinds
3. Market share to reach 40%
4. Export value for software and information technology to reach $0.4 billion U.S.

One notable fact is that, according to one report, by 1999 the total sales for the information technology industry had reached ¥174 billion ("Bright Future," 2000), more than three times as much as was projected. Another report puts the figure at ¥430 billion (Hou, 2000b). Regardless of which is the right figure, one thing is clear: the Chinese government is focusing its efforts on developing its strengths in software production so as to acquire a share in the international market.

The central role of the Chinese government in developing computer and Internet technology can be seen in its initiating efforts in developing the Internet in China. An article titled "Major events in the development of the Internet in China" published by China Internet Network Information Center (CNNIC) in 2000 records the series of major steps taken by the Chinese government in developing Internet technology. Following is an account of these major events, which provides a clearer picture of the pivotal role the Chinese government plays in technology development.

September 20, 1987: As mentioned earlier, Qian Tianbai sent out the first email from China to Karlsruhe University in Germany. That Qian was the first Chinese to do so was by no means an accident. A professor in Beijing Applied Computer Research Institute, Qian was in charge of the research project to connect Chinese Academic Network (CANET) to the Internet. Beijing Applied Computer Research Institute was one of the first research facilities representing the government to conduct such research.

1988: The High Energy Physic Institute of the Chinese Academy of Science connected its DECnet to the DECnet of the Western European Center as its extension, making it the first network to have international connection and making possible China's electronic mail communication with Europe and North America.

May 1989: China Research Network (CRN) successfully connected itself to the Germany Research Network (DFN). CRN's members included half a dozen universities and research institutes.

September 1989: The State Planning Commission opened its NCFC project to public bidding. NCFC was an advanced technology information project funded and organized by the State Planning Commission, the State Science Commission, the Chinese Academy of Science, the State Natural Science Foundation, and the State Education Commission. The purpose of the project was to build a super computer center and a high-speed network between three most well-known universities in China: Beijing University, Qinghua University, and the Chinese Academy of Science.

October 1990: Professor Qian Tianbai registered on behalf of China the top domain CN with the international Internet center in the U.S.

June 1992: At the INET92 conference in Japan, Professor Qian Hualin from the Chinese Academy of Science met with a key official from the U.S. National Science Foundation who was in charge of Internet administration to discuss the possibility of China joining the Internet. The latter told him that there would be political obstacles for China to be connected to the Internet due to the fact that many U.S. government agencies were online.

March 2, 1993: The High Energy Physics Institute of the Chinese Academy of Science connected its 64k line to the Stanford Linear Accelerator Center via AT&T's satellite. Upon its connection, the U.S. government limited it to the U.S. energy network alone, on the grounds that a socialist country should not have access to the science and technology information and other resources on the Internet. Nevertheless, this was China's first line to connect, though only partially, to the Internet. The State Foundation Commission provided generous support, including monetary funding, so that the leaders of some key research projects and hundreds of scientists were able to access the Internet and use email.

March 12, 1993: At a meeting over which he presided, Vice Premier Zhu Rongji unveiled a proposal and a plan to build a national public network for economic information communication, known as the Golden Bridge Project.

April 1993: The Computer Network Information Center at the Chinese Academy of Science organized a team of Internet experts in Beijing and conducted a survey of the domain systems of many countries. Based on the survey results, it determined the domain system for China.

June 1993: At the INET93 conference, network experts from China again expressed China's wish to join the Internet.

August 27, 1993: Premier Li Peng appropriated ¥3 million from the Premier Reserve Fund for the early phase of the Golden Bridge Project.

December, 1993: The State Economic Informationalization Joint Conference was founded, with Vice Premier Zou Jiahua as its chair. Also, in this same month, the main part of the NCFC network was completed.

January 1994: In the presence of the Joint Commission for China-U.S. Science and Technology Cooperation, the U.S. National Science Foundation approved China's request to connect NCFC to the Internet.

April 1994: The Joint Commission for China-U.S. Science and Technology Cooperation convened in Washington. At the meeting, Hu Qiheng, Vice President of the Chinese Academy of Science, reiterated China's request to connect to the Internet, which was approved by the U.S. National Science Foundation.

April 20, 1994: NCFC opened its 64k full-function connection via Sprint to the Internet. This marked the point when China was recognized as one of the countries having the Internet. This event was selected by the Chinese media as one of the top ten achievements of the year in science and technology.

May 15, 1994: The High Energy Physics Institute of the Chinese Academy of Science set up the first Web server in China and built China's first website.

May 21, 1994: With help from Professor Qian Tianbai and the Karlsruhe University in Germany, the Computer Network Information Center at the Chinese Academy of Science set up the server for China's top domain CN, thus ending China's history of having the CN server outside China.

June 8, 1994: The State Council General Office issued the following memorandum to all the ministries, commissions, and provinces: Circular of the State Council General Office on the Issues of the "Three Gold Project." This signaled the beginning of the full implementation of first phase of the Golden Bridge Project.

September, 1994: China Telecommunications and the U.S. Department of Commerce signed an agreement about the use of the Internet, which stipulates that China Telecommunications will open two

64k lines, one in Beijing and the other in Shanghai. China thus started the construction of CHINANET.

October 1994: China started the China Education and Research Network (CERNET), which was funded by the State Planning Commission and administered by the State Education Commission. The purpose of this network was to provide an information platform that would connect most of the universities and high schools in the country.

January 1995: As the two 64k lines set up in Beijing and Shanghai began to operate, China Telecommunications began to offer Internet service to the public.

April 1995: The Chinese Academy of Science began to expand its network beyond its own campuses to include research institutes in 24 cities. This network later developed into a national network called China Science and Technology Network (CSTNet).

May 1995: China Telecommunications began its preparation for building the mainframe network for CHINANET.

July 1995: CERNET opened its 128k line connection to the Internet in the U.S.

January 1996: The State Council Informationalization Leadership Group and its office were established, with vice premier Zou Jiahua as the group leader.

January 1996: The mainframe network for CHINANET was completed and began its operation. Internet service began to be offered all over the country.

July 1996: The State Council Informationalization Office organized a group of experts who surveyed China's four main networks and about 30 ISPs concerning their technical facilities and administration. This survey contributed to the standardization of Internet administration.

August 1996: The State Planning Commission approved the Phase I plan for the Golden Bridge Project and listed it as a key continuation project in the Ninth Five-Year Plan.

September 6, 1996: The China Golden Bridge Network (CHINAGBN) opened its 256k connection to the Internet in the U.S. and began to offer Internet service to the public.

December 1996: China's Public Multimedia Communication Network (169 Net) began its full operation.

April 18–21, 1997: The State Council held a national informationalization conference. This conference defined the meaning of a national informationalization system, its basic components, its guiding principles, its working principles, its major objectives, and its main tasks. The conference also drafted and passed The National Informationalization Plan for the Ninth Five-Year Period and the Objectives for 2000, which lists the Internet as a national information foundation project and announces its decision to establish a national Internet network information center and an Internet exchange center.

May 30, 1997: The State Council Informationalization Leadership Group issued *Temporary Administrative Measures on Internet Domain Registration in China,* which authorized the Chinese Academy of Science to establish and manage the China Internet Network Information Center (CNNIC) and authorized the China Education and Research Network (CERNET) and CNNIC to co-manage China's second-level domain ".edu.cn."

June 3, 1997: CNNIC was officially founded and began its operation. On the same day, the State Council Informationalization Leadership Group announced the formation of the CNNIC working committee.

1997: The four major networks, CINANET, CSTNET, CERNET, and CHINAGBN, were interconnected.

March 1998: The First Session of the Ninth National People's Congress approved the proposal to establish a new industry in China: information technology. This new industry would have under its jurisdiction electronic information product manufacturing, telecommunications, and software production. Its main function was to promote the informationalization of the national economy and social service.

July 1998: China Information Safety Testing and Examination Center passed inspection by the State Council Informationalization Leadership Group and began its test operation.

July 1998: Phase II of the mainframe network project for CHINANET began its construction.

January 1999: The mainframe satellite network for CERNET completed its construction. In the same month, CSTNET opened two satellite systems. Both projects greatly increased the access speed of China's Internet.

February 1999: China National Information Safety Testing and Examination Center (CNISTEC) began its formal operation.

As is obvious, the Chinese government was the main initiating and activating agent in these events that mark the development of China's Internet. Indeed, the government determined the development path of China's Internet. Although some of the agents involved are research institutions and universities, they all represent the government in one way or another since, without exception, they are all funded by the government.

It is no exaggeration to say that the government's role in the development of this new technology is omnipresent, and there is no indication that the government is ready to relinquish or even diminish this role. Instead, its leaders have been consistently promoting new technology on all occasions. When inspecting high-tech development projects in Beijing earlier this year, for example, Jiang Zemin, the President of China, said that "China should make renewed efforts to develop high-tech industries and step up technological innovation, which is the driving force behind social and economic progress" (Dian, 2000). Such an orientation was reiterated by Gao Xiqing, vice chairman of the China Securities Regulatory Commission, who said that "the government would create favourable market conditions for the so-called new economy," which "refers to rising industries such as e-commerce, telecommunications, computers and information technology" ("Stocks cheered," 2000). In a talk at the Asia Society 11th Annual Corporate Conference, Premier Zhu Rongji reassured more than 1,000 international business delegates, government officials, and scholars of China's determination to speed up the country's opening process to improve the economy (Xie & Zhang, 2000).

In speeding up the opening process, computer and Internet technology will certainly play a significant role, defined and realized to a large measure by the government. In fact it already is. Under the strategic guidance of the Chinese government, computer and Internet technology is pervading many areas, old and new, some of which include boosting regional economy (N. Cui, 2000c; Zhu & Tan, 2000; Z. Chen, 2000; Huo, 2000b; L. Chen, 2000; Huo, 2000a), providing new technological impetus for many fields and industries (M. Sun, 2000; Zeng, 2000; Liang, 2000; Hou, 2000c; Cao, 2000), spearheading e-commerce (X. Xu, 2000; J. Lu, 2000; Wang & Cheng, 2000; Xue, 2000; Smith, 2000), and technologizing education (N. Cui, 2000a; N. Cui, 2000b; Xiao, 2000; Feng, 2000; J. Cui, 2000,).

The resolve to develop advanced technologies including computer and Internet technology has never been so strongly emphasized by Chinese leaders, and rarely has the government been so eager to play, and has played, such a major role. One reason behind such eagerness may have been the monopolistic control characteristic of almost all the governments in Chinese history in its major industries. As Huang (1990) points out, "despite the recent decentralization, the Chinese government is never inclined to relinquish its monopolistic control of metallurgical, petroleum, and chemical industries, machine-tool production, shipbuilding, public transportation, banking and insurance, foreign trade, broadcasting, and even the tourist industry" (p. 254–255). The IT industry is certainly no exception, especially when computer and Internet technology is of such critical importance to almost all the other industries.

Another reason, as we pointed out earlier in this chapter, is that improving the economy has become a necessary condition rather than a luxury option for maintaining Communist rule in the last couple of decades. Therefore, instead of passively being pushed into technological change and running the risk of losing control over its consequences, the Chinese government has decided to be an active agent in this change process so that it can shape its development path, although the extent to which it can shape the path is not totally in its control.

Yue-Farn Wang (1993) attributes the Chinese government's active participation in making science and technology policies to a different motive: its pursuit of international prestige:

> It seems that throughout the forty years of PRC history, a large proportion of both state and scientific ac-

> tors have tended to use science policy for the purpose of pursuing international prestige rather than pursuing China's concrete development needs. *Apart from the efforts of certain government reformers in the 1980s to press for social relevance, the interests (individual or group) of Party leaders and the research communities prevailed, leading them to continue to disregard social accountability.* (p. 147).

The personal interests of the non-political-elite individuals (such as scientists) notwithstanding, Wang is certainly right in arguing that the Party leaders (and thus the government) are more interested in Party interests than in social accountability. However, international prestige alone seems a rather oversimplified explanation as, obviously, maintaining the Party's control over the state necessarily precedes seeking international prestige. In addition, putting the Party's interests before the people's is not always incompatible with social accountability. This does not mean that the Communist government in China is a conscientious one seeking the welfare of its people. Rather, it means that in trying to achieve its ends (to stabilize the society and thus to consolidate its rule), social accountability (for example, improving the economy and thus people's living standards) sometimes becomes a necessary means. Inadvertently, then, the pursuit of Party and individual interests by the Communist government sometimes leads to actions that are at least seemingly conducive to social accountability. Given its number-one priority, the Chinese government, therefore, is understandably active in this new change process.

Granted that even without this active participation by the Chinese government the development of computer and Internet technology in China would have been inevitable, it is impressive that it developed at such a fast pace. Regardless of the motives behind this active participation, objectively, the active role assumed by the Chinese government has produced a positive effect on the development process. However, such an active participation by the government is not without its costs. While the government recognizes the significance of the development of the technology, and thus the economy, to the safeguarding of its hegemony, having total control over this development and its consequences so that they do not threaten its reign has always been the number-one priority. Thus, along with its efforts to promote this new technology, the Chinese government has also been consistently making an effort in the opposite direction: to control the "negative" (mostly ideological) consequences of this new technological development. This effort is reflected in the series of laws and regulations made by the Chinese government concerning the development of computer and Internet technology. A look at such laws and regulations on Internet development over the last few years clearly illustrates my point.

Corresponding to the Internet development in China since the mid-1990s, the Chinese government has established a series of regulations related to Internet development (see Table 3 for a list of the major regulations). The purpose of these regulations is two-fold. On the one hand, as this new technology develops at a tremendous speed, there is a pressing need for laws to regulate this new technology. The Chinese government is well aware of the risks of uncontrolled, unregulated development. On the other hand, an equally important purpose of these regulations is to control the general public's access to information on the Internet. And there is reason to believe that this second purpose in many ways outweighs the first.

Table 3. Laws and Regulations on Internet Development in PRC

Date of Issuance	The Regulation
2–11–1996	Temporary Provisions for the Management of Computer Information Networks in the People's Republic of China That Participate in International Internet Systems
5–20–1997	State Council's Decision on Revising the "Temporary Provisions for the Management of Computer Information Networks in the People's Republic of China That Participate in International Internet Systems"
12–8–1997	Implementation Measures Relating to the Temporary Provisions for the Management of Computer Information Networks in the People's Republic of China That Participate in International Internet Systems
12–30–1997	Administrative Measures on the Safety/Protection of Computer Information Networks that Participate in International Internet Systems

Not available	Regulations on the Safety/Protection of Computer Information Networks in the People's Republic of China
Not available	Temporary Measures on the Administration of the Registration of Internet Domain Names in China
Not available	Detailed Regulations on the Implementation of the Registration of Internet Domain Names in China
1–25–2000	Regulations on the Security Administration of Computer Information Networks That Participate in International Internet Systems

Ever since the Internet became available to the general public in China, along with access to this new technology came access to a wide array of information, including capitalist ideologies that directly challenge Communist doctrines. The Chinese government is acutely aware that with every new technology comes new ideas, "good" and "bad." The Internet has brought with it unlimited access to all the information that the Communist government has feared the most and has made every effort to block from the masses. The potential influence of capitalist and democratic ideologies on the people's way of thinking is the "worst" trade-off of this new technology. As Green (2000) points out, "The Internet could be an important weapon for undermining regimes which lack a democratic mandate. With eight million online and counting, China's rapid Internet development should be a major threat to the political order."

To diminish this threat, the government has issued strict policies that restrict Internet access, including those major regulations mentioned above. The intended functions of these strict rules range from regulating the general use of the Internet to controlling online trade and e-commerce, to restricting the use of foreign-designed encryption technology, and to even tightening control on e-cafes (see, for example, Chu, 2000; Forney, 2000b; Sui, 2000; L. Xu, 2000; Y. Zhang, 2000). These policies and regulations have produced a negative, sometimes rather detrimental, effect on Internet development in China.

Thus, it seems, the Chinese government has been playing a dual role of Jekyll and Hyde in the development of computer and Internet technology. Indeed it has. The main reason for its playing such a dual role is that the government is always in a mindset of wanting to eat the cake and have it, too. As in the case of any technology transfer, the government has to constantly balance between enjoying the benefits of the new, foreign technology and suffering from its consequences. The result is a series of seeming self-contradicting steps and measures taken by the government. However, one thing must be pointed out. The guiding principle for the Chinese government in developing computer and Internet technology seems to be to promote development to the maximum and to keep restricting measures to the minimum, provided that the development does not pose a serious threat to its rule. Obviously, the development of this new technology, on the whole, seems to benefit more than undermine its hegemony. Hence, we have seen this dominating participant role played by the Chinese government.

The Masses—The Subtle Contributor. With the government playing such a dominating role, a logical inference would lead to the conclusion of a dismissible role for the general public in this development process. The truth, however, proved to be far more complicated than such a convenient conclusion. The general public, in fact, has contributed to this development in a significant, though rather subtle, way.

As mentioned earlier, generally, there are three major groups of participants in a technological development in a society: the political elite, the technical elite, and the general public. We have discussed above that in the China case the role of the political elite is mainly played by the government. While the political elite has the power to make policy and managerial decisions, the technical elite is the group that possesses the technical knowledge of the technology. In many ways, the technical elite influences the political elite in policy-making and the users (the general public) in understanding the technology. In the case of computer and Internet development in China, the technical elite group does exist. However, with the government playing an unusually strong role in this technology development, the technical elite group loses, to a considerable degree, its power to influence policy making, thus blurring its distinction from the general public. This is due to several reasons.

First, as mentioned earlier, the Communist government has always been unwilling to relinquish its power to control almost anything, even to the smallest degree. When compared with the knowledge power the technical elite possesses, the ignorance of the political elite (in this case the government itself) concerning the technical aspects puts the government at a distinct disadvantage when it comes to defining the technical meaning of that technology. One thing the government tries to prevent is the intensification of that disadvantage. Therefore, it will do anything to minimize the knowledge power of the technical elite. This is reflected partly in its efforts to exclude, as much as possible, the technical elite from the decision-making process. This does not mean, however, that the technical experts are not involved at all in decision-making. In fact, many technical experts do participate in the process. However, in their participation, these experts play a dual role. Often, these experts are the ones who also occupy high managerial positions. For example, the president of the Chinese Academy of Sciences, which has been quite active in the development of computer and Internet technology, is as much a political position as an academic one; in fact, in many ways it is more political than academic. These technical experts then belong to both the political elite and the technical elite. This dual role requires them to interchange between two functional capacities. When they participate in decision and policy making, they are functioning as part of the political elite, rather than as members of the technical elite.

Second, according to Rensselaer Lee (1977), the Communist government has always sought to eliminate, or at least contain, social stratification and to "mobilize mass participation in the technical sphere" (p. 289). In the earlier days of the Communist rule, such an orientation resulted in "a kind of inverted Marxism that legitimizes the resettlement of urban dwellers in the countryside and, most important, regular participation in manual labor for intellectual elites" (p. 289). Today, though the trend is toward urbanization, the integration of the intellectuals and the masses is still a much emphasized point, at least on the part of the government. In theory, the elimination of the social distinctions between mental and manual labor works in the favor of the government because inequalities are always a potential source of social unrest. However, how successful the government is in this effort in today's China is questionable.

Third, computer experts in China are not likely to replace part of the political elite—the middle-level political leaders—in managing the technological development at even the local level. This group of middle-level political leaders, though not a powerful sector within the political elite group, is the backbone of fundamental support for the Communist government at local levels. Over the years, they have proved that they are a reliable source of support for the Communist

Party. The technical elite, on the other hand, is a newly formed group, with many qualities still unknown to the Communist leaders. The Communist Party is not likely to sacrifice the support of local leaders and entrust their local rule to this unknown group of technical experts. Instead of turning the decision-making power over to the technical elite, the Communist government would hand select certain members from this group and turn them into part of the political elite. The special privileges accorded to these new members, and the fear of losing such privileges, are often effective ways of ensuring their loyalty to the Party and the Communists' control of the decision-making power.

Fourth, even though there are now many computer experts in China and the number will only grow, it is doubtful whether these computer experts will form a cohesive technical elite group. They lack a fundamental adhesive that could piece the group together, a clear identity and interest that every member of the group could identify with other than the fact that they all possess computer expertise. Therefore, they are more a loosely formed group than a well-knit whole. Though, undoubtedly, they are actively participating in and contributing to computer and Internet technology development, their involvement is not unique enough to differentiate them much from ordinary users.

How, then, do ordinary users (the general public/the masses) contribute to this development process? To answer this question, it is necessary to review the nature of the technological reform that the computer and the Internet have brought to China. In a free society, technological change is a participatory process involving everybody who is affected by the technology. All the participants define and redefine the technology by contributing and negotiating their perspectives. In a totalitarian society, however, although the change process still involves all the participants, the form of contribution and negotiation is fundamentally different because the nature of the hegemony and the dominating participation by the government determines that the technological reform is a top-down process. This inevitably leads to what Feenberg (1991) calls "the *paradox of reform from above:* since technology is not neutral but fundamentally biased toward a particular hegemony, all action undertaken within its framework tends to reproduce that hegemony" (p. 65, emphasis original). This does not leave much room for the general public, the less powerful participants in a socialist system like China's, to create a new social organization that is different from the present hegemony. How, then, do the masses in China contribute to this process of technological change?

The masses in China, as the ordinary users of the technology, play their role in two subtle ways: playing by the government's rules of the game and playing against them. First, playing by the government's rules of the game, the ordinary users maneuver within their limited margin to influence the development in their favor. As mentioned above, the Communist government, from the time of Mao to those of Deng and Jiang, has always promoted mass participation in technological innovations. The proclaimed effort to eliminate social stratification, in effect, works in the general public's favor in that some of the power that originally belonged to the technical elite is shifted to the ordinary users. As the government issues various policies to promote computer and Internet technology development, the general public is often quick to seize the opportunities allowed within the legal limits of those policies. A pattern of remedial policy making by the Chinese government has also benefited the general public in this respect. Often, in haste to promote the technology, the government would make some policies that are extremely generous and lenient in giving people freedom. These policies often contain loopholes, which the public is quick to take advantage of. By the time the government realizes that these loopholes have produced consequences far beyond its expectations, it is often too late to reverse them, even though they would then make some remedial policies in hopes of "correcting" the course. In the mid-1990s, for example, when Internet development was at its still rather early stage of development, the government issued some policies that allowed rather free access to the Internet. When it was realized that people were accessing information that they were "not supposed" to access, the government released several policies in an effort to limit this access, but it was clearly too late to undo the "consequences."

Besides playing within the rules of the game, a second way the ordinary users play their role is to directly resist and even alter some of the rules. As mentioned earlier, most of the official regulations concerning Internet use issued by the Chinese government since the mid-1990s are designed more to limit people's access and mend the loopholes than to regulate the industry. While some of the regulations are generally heeded by the general public, some others have been resisted by the users.

A report by Henry Chu (2000) in *Los Angeles Times,* for example, reports that despite China's "latest series of rules regulating Internet use" including "making it a crime to leak 'state secrets' electronically, requiring companies to register their sensitive encryption software and warning 'dot-coms' against hiring reporters to provide news not approved by the government . . . Internet and computer firms are continuing with business as usual." According to Matt Forney's (2000b) report, "companies in China had until yesterday to tell the government what software they use to send secure information over the Internet, but the deadline passed with many companies still unclear how to register and others refusing outright." The reason, according to some Internet users, is that they either expect the rules to weaken or do not believe that the rules will be enforced.

This refusal by companies to comply with the rules is more than incidental. In fact, many such rules and regulations have met with resistance from users. There are two reasons behind this resistance. One is that, up to the present day, China has never been a law-governed country. Instead, it has always been person-governed. In other words, power has never rested in the laws or even in the official positions. Rather, it has always rested the person making the laws or occupying the positions. A good example of this is that Deng Xiaoping remained the actual paramount leader of China in his last years of life even though he relinquished his official positions in the Party and the government. What this means is that the laws, at least some of them, do not have much binding power. The general public has learned over the years not to interpret regulations literally. As Forney (2000b) describes, "some foreign companies have chosen to interpret the regulations loosely." Others tried to guess what the government really wants: "We all try to comply with the regulations, but when you're not sure what they're trying to get at, it's hard to comply" (Chu, 2000). This open resistance sometimes results in the government giving up on the policy and thus the reinforcement of the status quo. An example is that "a policy that all Internet users register with the police, a requirement greeted with alarm when it was issued some time ago, has gone virtually ignored" (Chu).

Another reason for this resistance is that the government has exhibited a pattern of half-heartedness in following up on their regulations. Though some of the regulations are enforced strictly, many others appear to be items on the government's wish list and are often not followed through sincerely. Many such regulations end up either ignored or sometimes even reversed. Last year, for example, "China banned foreigners from investing in the Internet but then realized its Internet companies would crash without funding from overseas" (Sui, 2000). In the early 1980s, China required all foreign companies to register all their telephones and fax machines. That rule was also ignored and later rescinded (Sui). Another regulation put a lot of restrictions on the more than 1,000 e-cafés in Beijing, "but few adhere to the proper procedures and regulations" (L. Xu, 2000). As for the rule for the foreign-design encryption software, it was reversed in March, only a month after it was issued (Forney, 2000a).

Although the reversal of such rules and regulations is partly due to the fact that they are not well thought out by the government and lead to unforeseen ramifications that might suffocate technological development, it is also a result of open resistance by the general public. By playing both by the rules of the game and against the rules within reasonable limits, the masses can successfully increase the margin of maneuver and alter the rules in their favor. Although they may not be in a position to replace the Communist technical code in this instance, at least they can subtly influence it to such an extent that it would, in Feenberg's (1991) terms, "take into account a wider range of variables" (p. 158). In other words, the Communist code of technological development does not have to be exclusively reflective of the ruling class code only. Instead it could incorporate non-ruling-class variables and reflect to some extent the interests of the general public as long as the public played the game wisely. Even within a non-participatory system of technological development like China's, there is always room for every participant to play his/her appropriate role. Therefore, in examining computer and Internet technology development in China, the role of the general public is not to be discounted.

Knowledge Creation—Variously Constructed
Realities about Technology

As was mentioned in Chapter 3, technological development, as a process of knowledge construction by various participants, is a cultural act that takes place at both the individual and the collective level. The development of computer and Internet technology is certainly no exception. As discussed earlier in this chapter, there are mainly two groups of participants in this development process: the political elite (i.e., the government) and the general public (i.e., the ordinary users), although each group has its subgroups. With its dominating role, the government has been an active initiator in defining the meaning, the purposes, and the uses of this technology.

The role of the Chinese government in defining the meaning of computer and Internet technology is reflected mainly in its efforts to enforce its Communist technical code through various means. Equating technological rationality with political rationality, Feenberg (1991) points out that "the values and interests of ruling classes and elites are installed in the very design of rational procedures and machines even before these are assigned a goal" (p. 14). He further argues that the dominant form of technological rationality "stands at the intersection between ideology and technique where the two come together to control human beings and resources in conformity with what I will call 'technical codes'" (p. 14).

If this is true, and it cannot really be otherwise, in a hegemonic system like China's, the enforcing of a Communist technical code is thus inevitable. Such a technical code is reflected in the series of policies, laws, and regulations that define the parameters of this development. If we accept Feenberg's concept of technological ambivalence, we can see that computer and Internet technology can be defined in at least two different ways: positive or negative (if we could use these overly simplistic terms for the moment). Though the Chinese government has defined this technology in both ways on different occasions, it has manipulated, to say the least, its almost exclusive power to define this technology to serve the purposes of its communist regime. As mentioned earlier in this chapter, the development of this new technology is an important means to economic development, which in turn is a means to achieving social stability and thus consolidating the Communist rule. In presenting this technology, however, the government has portrayed economic development as the end and computer and Internet technology as an important means for this end. Defined this way, the purpose of this new technology then, seemingly, becomes the improvement of the welfare of the people. In some way, the technology does serve this function, but as discussed earlier in this chapter, social accountability is not what the government is concerned about.

At other times, though less frequently, the Chinese government has also portrayed this technology as possessing evil aspects, especially its ability to disseminate information that could "corrode your soul," such as pornography, or "endanger national security," such as state secrets. Hence we have the various rules and regulations aimed at limiting people's access to information. Although it is doubtful that the general public would buy into such images of the technology, in effect, it does limit the parameters of definition by limiting its uses through "legal" means. Overall, the government has portrayed this technology in a positive way, and our discussion of the government's role in this development earlier in this chapter should provide testament to this assertion.

The general public, on the other hand, defines this technology quite differently. For most ordinary users, the new technology represents new knowledge, new skills, which could lead to better jobs and improved financial and social status. In a country where political or financial opportunities used to be scarce, this new technology, especially the Internet, in many ways is a new way of life. Wang Siping, a woman who never attended high school and who came from Qinghai Province, one of the poorest regions in China, found a job through the Internet and now works in the southern metropolis Nanjing as a secretary for a Hongkong business man (Pomfret, 2000). "If it wasn't for the Web, I'd still be back in the desert," she said (Pomfret). For many, the Internet is a means for financial self-actualization.

For some others, the Internet is a means for cultural and political self-cultivation. For Sun Lingsheng, a music student at the prestigious China Academy of Opera and Dance, for example, "the Internet is becoming the great equalizer, within and without China," a ladder to climb the walls between people, class walls, cultural walls, and national walls (Platt, 2000). Cyberkiller, a young hacker in Beijing, shares the same perspective. For him, "Internet is not only a peaceful promoter of pop culture, but also a potent weapon," "a slingshot that can be used to slay traditional Goliaths" (Platt).

For still others, the Internet, as a new way of life, has a different meaning: it is changing some age-old cultural traditions. For many Internet users, traditional greeting cards sent around the Spring Festival, for example, are now replaced by electronic cards (Guo, 2000). Some go much further. Wang Hao, a 17-year-old Internet user in Beijing, "broke with the ancient Chinese custom of making dumplings on New Year's Eve" and hired, over the Internet, one of Beijing's most famous Sichuan-style cooks to cook his family a spicy meal in the family kitchen for a modest $19 (Guo).

Perhaps, one of the biggest changes computer and Internet technology represents and has brought about is the change in the knowledge structure in Chinese culture. In this ancient culture, where respect for the elder is always emphasized and where age represents knowledge and wisdom, this technology may be bringing about a change as there is a tendency to reverse such a perspective. The older people are being left out or left behind in this new development where, for example, over 75 percent of the Internet users are under the age of 30 (S. Sun, 2000). It is no longer an unusual thing for the parent to learn from the child about computer and Internet technology. Mao (2000) has this account:

> "You are wrong! My son told me that . . ." blurted out Zhou Xian, a professor from the Chinese Department of Nanjing University during a heated discussion with several of his colleagues over a computer problem.

By saying that, Zhou no longer put himself in the position of a senior or the authoritative person in front of his son. It is now very common that Chinese parents are willing to be the students of their children in learning some modern things such as computers.

Such a change concerns more than knowledge structure. The capability of the computer and the Internet to represent and present knowledge is effecting a "cultural feedback" that is affecting "a wide range of fields, from values, life attitude, formation of social behavior models to knowledge and use of new implements" (Mao, 2000). An important consequence of this cultural feedback is the way people perceive knowledge, its construction, its access, its possession. Computer and Internet technology represents both knowledge itself and the source of knowledge.

Obviously, constructions of the meaning of the new technology differ from group to group (government vs. the general public) and from person to person (among the members of the same group). The incompatibility between the government and the masses in their perspectives about computer and Internet technology makes it difficult to construct meaning at the collective level. Nevertheless, such collective meaning construction is inevitable, as it is a prerequisite to technological development. What results thus is a constant negotiation between the two participant groups, leading to a compromise between the groups in terms of their participant roles and definitions of the technology. On the one hand, the general public recognizes the dominant role of the government in mapping out the parameters of the technology and is careful not to cross the boundaries. This results in the government defining the nature of the development, its main purposes, and its main uses. On the other hand, the ordinary users contribute to meaning construction by taking advantage of their limited power to interpret and redefine, within their allowed margin of maneuver, the parameters in specific contexts. Hence we have the constant rule making by the government on the one hand and not infrequent resistance by the masses on the other.

Knowledge Access and Control—Pluralized Ownership

Discussion of knowledge construction begs an inevitable question: who owns the knowledge? Although my above discussions have rendered it quite transparent as to who has power in defining that knowledge, the access and control issue is never a simple one and thus defies simplistic generalizations.

Computer and Internet technology is an important means to all participants regardless of their interests and ultimate ends. Ideally, knowledge and its control exist in three loci: political, intellectual, and practical. The political locus is where the policy-making power resides and where ideological mapping of development takes place. The intellectual locus is where technical knowledge exists. The practical locus also possesses knowledge power in that it determines the way the knowledge is used. These three loci correspond to the three participant groups that are usually present in technological development: the political elite, the technical elite, and the general public. As I pointed out in Chapter 3, possession of and access to knowledge are not equal among participants. This is especially true in a social regime like China's. In fact, as discussed earlier in this chapter, the distribution of power in the China case is greatly unbalanced.

High on the power continuum, the Chinese government as the political elite controls the ideological mapping of the technological development. Its power, however, does not stop at this level. As mentioned earlier in this chapter, it also takes over much of the power that usually belongs to the technical elite by including members of that group in the political elite group. The usual ambivalent role the technical elite plays (i.e., a hybrid of the roles of the political elite and the masses, is somewhat lost on the technical experts in China. What results is an empowered political elite group and a rather disempowered technical elite, which blends into the general public group.

At the other end of the power continuum, the general public enjoys little power in defining the technology. The seeming lack of social stratification should mean, in Marxist perspective, a more empowered general public. The result, however, is the opposite. For one thing, the elimination of social classes is more an ideal than a reality in present-day China. For another, the polarization of participant groups into the government vs. the general public exacerbates the problem. When we have the three groups in their normal functional capacities, the power is more or less evenly distributed and balanced. In the China case, however, the concentration of power by the political elite renders the general public much less powerful than it should be. The ordinary users are left with a limited margin of maneuver.

The selection of an appropriate term for *Internet* described in Sheng He's (2000) report in *China Daily*, "New tech words stir debate," in some way provides a good example of knowledge construction and control in China. Currently in China, there are three three-character terms used to mean *Internet: Yin Te Wang, Hu Lian Wang,* and *Wan Wei Wang*. These three terms are constructed according to different methods. *Yin Te Wang* is a combination of sound and meaning translation. *Yin Te* is the Chinese pronunciation equivalent to "inter" while *Wang* is the direct translation of "net." The three characters combined do not really have meaning in Chinese. *Hu Lian Wang*, probably the most widely used term of the three, can be loosely translated to "interconnected net." *Wan Wei Wang*, a term mostly used in academic circles, is a the result of an attempt to preserve the three Ws in the English term the World Wide Web). *Wan* means "ten thousand," a popular term to mean many, *Wei* means "dimensions," and *Wang* means "net."

All three terms are recognized by Internet users in China. However, the government's intention to unify the terms for fear of legal complications in the future resulted in a competition of definitions: "the three expressions for the Internet have vied for official recognition since the Internet was first introduced to China in the early 1990s," and "a heated debate raged and countless translations were suggested" (He, 2000). Many people, "scholars, students, IT technologists and even poets," joined the debate (He). Logically, such a controversy would end when one of the terms edges out the other two and wins people's preference. In China, this was not the case. The issue was "settled" by the intervention of the China National Committee for Terms in Science and Technologies (CNCTST) when it selected *Yin Te* Wang as the official term. CNCTST was authorized by the State Council to approve the official use of new words. It decided that *Yin Te Wang* would be used to refer to the Internet and that *Hu Lian Wang*, the purely Chinese word, would be used to refer to the broader concept of the Internet. Though some supported this decision, "others dismissed the terming as confusing and a violation of the tradition of the Chinese language," and still others described the term *Yin Te Wang* as an "uninteresting," "lifeless," and "not user-friendly" term "that apparently assumes the arrogance of the [bureaucratic] intellectuals"(He).

That different people prefer different terms is not surprising. Some terms will eventually survive others. In the case of China, however, the power of administrative intervention can often change the course of natural development. Prescribing language use, a practice that has already been abandoned in the West, is still amazingly active in China. As a result, "use of the hybrid term is now spreading quickly with the aid of the "administrative power" (He, 2000). The same thing happened with the translations of the term "ergonomics" earlier. These are more than simple cases of prescribing language use. Instead, they illustrate a totalitarian way of defining terms and constructing meaning in China's Communist regime that is decided by the hegemonic hierarchy of the power structure.

The uneven distribution of power in China, however, has not led to a loss of balance in the social order. This is due to two reasons. First, as I mentioned in Chapter 3, no participant group has absolute power, even under a socialist system. Even though there are, in effect, only two large, distinctive groups, the dominant power of the Chinese government is relative: without the general public, the government's power would be meaningless, and therefore, there would, in fact, be no power. This forces both groups to acknowledge the role of the other group and give each other room to maneuver. Second, the discrepancies in the level of control over knowledge about computer and Internet technology between the government and the general public, and among their subgroups, make it necessary for these groups to cooperate with one another and to combine their strengths for the common good that will benefit all the parties involved. Nevertheless, the hegemonic power structure in knowledge construction in China will have adverse effects on the development of computer and Internet technology in the long run.

The Communication Medium—A Multiplicity of Distribution Means

While one may rightfully claim that the medium for knowledge distribution is an inseparable part of knowledge construction and access and should be examined in close conjunction with the latter two, it is singled out in this study as a topic in its own right for two reasons. One is that the medium used for knowledge distribution often exerts a significant impact on the shape of the outcome of that knowledge being distributed. Another, less obvious, reason lies in the unique case of the communication media in China. In Western, democratic cultures, communication media are various and are generally accessible to the general public, even though this access is to varying degrees with different users. In China, however, while most, if not all, conventional media are existent, they are far from accessible to the masses. The government's monopoly and manipulation of the media and the general public's subtle fight for access to the media therefore make a noteworthy topic in our examination of the ways these two participant groups attempt to shape the construction of knowledge about computer and Internet technology.

Two aspects about the communication medium will be discussed in this section: through what medium knowledge about computer and Internet technology is communicated in China, and how the computer functions as a new writing medium in Chinese. These two seemingly unrelated aspects address two important issues of communicating about and through the computer. While the first aspect addresses the important question of the development process and the form of rhetorical construction, the second aspect discusses the development potential of this new medium in the Chinese context.

China presents a unique case in terms of the medium for distributing and communicating knowledge about computer and Internet technology. As I mentioned in Chapter 3, public access to technical knowledge in a socialist society requires four mechanisms: a reward structure that encourages public access, an organizational structure that gives the power to the masses, a system that discourages technocratic and encourages democratic attitudes, and a formal means of intervention by the workers. In China, none of these mechanisms are exactly in place, although the government has made certain limited gestures toward the democratization of knowledge access and distribution. This results in the government's attempt to disseminate technical knowledge and encourage public access while making sure to have control of the medium for communication. Under such an orientation, the official media of the government—newspaper, radio, TV, etc.—inevitably becomes the main, sometimes the only, media for communication.

To understand the magnitude of the government's control of the communication media, one must know that a majority of the mainstream media in China are controlled by the government, central or provincial. As Wu (1994) points out in his analysis of the politics of editorial formulation in the *People's Daily*, the official newspaper of the government, "mass media and the party-state are seen as identical in essence, as implied in the concept of 'propaganda state,' [which] restructures people's opinions and transforms society" (p. 194). Such an assessment is no exaggeration. Although it is beyond the scope of this study and also impossible to do a comprehensive analysis of the role all the media in China has played in the dissemination of knowledge about computer and Internet technology, a partial examination of *People's Daily* should still provide us with an insightful glimpse into the manipulative operations of this propaganda state.

What I will do next is to take a look at the news reports about computer and especially Internet development that appeared in the *People's Daily* in roughly the first half of 1997. The reason I chose *People's Daily* is that, as I explained, it is the official newspaper of the government. It "is the mouthpiece of the Central Committee of the Chiense Communist Party . . . , the top decision-making body in China, and is controlled by the Propaganda Department of the Central Committee" (Wu, 1994, p. 195). Funded and controlled in every aspect by the government, it has also been the most widely circulated newspaper in the country. It is, in a way, the "spokesperson" of the government. As for the historical period, I chose mainly the first half of 1997 because the Internet started in China in the mid-1990s. From early 1997, China was speeding up the pace of Internet development. This period represented a segment of the developmental stage for this new technology. Both the government and the general public were in the stage of getting to understand the functions and capabilities of the Internet. In a sense, it is a time when they were in the middle of forming their perspectives on this new technology. Therefore, it is interesting to see what perspective the government communicated about the technology and how it communicated it.

An examination of computer and Internet related news reports that appeared in *People's Daily* during this period revealed that the editors strategically selected the type of reports to publish. As mentioned earlier in this chapter, the government recognized the importance of the development of this new technology to the consolidation of its hegemony; the overall tone of the reports clearly reflected this orientation as they, without exception, promoted the positive aspects of the new technology. Among others, several themes were obvious. One was the persistent reports about new developments in computer and Internet technology, creating the impression of a fast development. A second theme was the frequent column articles that provided consultation and free information service on Internet use, encouraging the general public to use the Internet. A third theme was its strategically constructed reports about the penetration of this new technology into different fields, projecting an image of pervasiveness and ultimately encouraging all fields and disciplines to use the new technology.

Let's take a look at this third theme, how the *People's Daily* projected the image of the penetration of Internet technology into all walks of life. Following is a summary of the reports in the chronological order, with each presented in the order of its date, the title of the report, and a brief summary of the news.

December 28, 1996: "The Internet knocks on the doors of the ordinary people." Several major department stores and shopping centers offer a new product, which is an Internet package that includes the equipment and information needed for Internet connection.

January 27, 1997: "'Around China' attracts international tourists for you through the Internet." A travel agency opens China's first Internet site on tourism, which will provide reservation and other services for hotels, restaurants, transportation companies, etc.

March 20, 1997: "First contemporary Chinese novel up on the Internet." The Chinese novel *Key* goes on line.

April 19, 1997: "'Internet Building' emerges in Beijing." Beijing's Saite Building, an office building, now provides Internet connection and Internet related services to the companies that rent space in this building.

May 19, 1997: *"City on the Internet:* A window to Fujian." A website in Fujian provides comprehensive information about real estate in Fujian.

May 21, 1997: "'Going Online' getting hot on both sides of Huangpu River." Going online is becoming very popular in Shanghai.

May 26, 1997: "'Golden Talents'" site to provide employment service." "Golden Talents," a government funded project, sets up a website to provide employment-related services.

June 9, 1997: "A culturally Chinese 'Read Online.'" "Read Online" is a site that provides a wealth of information related to the Chinese culture.

June 13, 1997: "Internet café." The first-ever Internet café opens in Beijing.

June 17, 1997: "Chinese basketball enters the Internet." China Basketball Association opens its website on the Internet; basketball fans now can access all kinds of basketball information through the site.

July 7, 1997: "China Export Commodity Fair on the Internet." The United Nations Trade Network China Development Centers opens a website for China Export Commodity Fair.

July 10, 1997: "Bank of China releases financial news on the Internet." Bank of China now releases financial service information through its website.

July 17, 1997: "Wuliangye recovers its international domain name." Wuliangye, a liquor manufacturer succeeded in retrieving its domain name "Wuliangye.com" from a U.S. information company.

July 21, 1997: "China Finance & Stock Investment Information Network." Hexun Company, a subsidiary of China Stock Market Research and Design Center, opens an exclusive website on stock exchange.

These were only some of the news reports related to computer and Internet development published in *People's Daily* during this period. A notable feature about these reports is that the disciplines they represented were all different, with no two reports about the same discipline. The strategic orientation to promote Internet use was evident and consistent with the government's overall Internet development policies at the time. Even though some of the news reports may not seem to be significant enough to warrant publication in a nationally circulated official newspaper, they were compatible with the newspaper's overall strategies.

Perhaps, two computer-related, but not Internet-related, reports published in *People's Daily* on the same day, on January 2, 1998 are more revealing of this orientation. Both reports were about the use of computer technology by farmers. The first report, "Young woman uses computer to raise chickens and makes ¥20, 000 a year," tells of a young farmer who greatly increased her income by using computer technology in raising chickens (Li & Li, 1998). A second report, titled "Computer instructs farming," is rather farfetched in bringing the computer into the topic. It tells of an old farmer, who had been illiterate all his life, who encountered some problems with his apple trees. He therefore went to the local farming technician for help; the technician was able to retrieve some useful information on the computer, which helped the farmer to cure the disease in the apple tree (Li & Li, 1998). These two reports touched a new subject, the use of computer technology in the traditionally underdeveloped rural areas. In some way, they illustrate the government's resolve to promote this new technology in all disciplines.

Given such monopolistic control of the distribution media, is there any room in the official media for ordinary users to communicate their perspectives? There is, but in a very limited sense and only occasionally. One example illustrates the limited margin of maneuver for the ordinary user. From April to October 1998, *People's Daily* ran a series of eight articles written by a well known writer Wu Yue about what some of the famous writers in China went through in computerizing their writing medium (i.e., in switching from the pen to the computer). This series is worth mentioning not only for the reason stated above but also because it touches the second aspect about the communication medium: the development potential of the computer as a new writing medium and the effect of the unique characteristics of the Chinese language on such a potential.

The context behind this series was that from the early to mid-1990s, many writers in China were going through a transitional period during which they were switching their writing medium from the pen to the computer. By 1998, most writers had adopted the new medium. Wu Yue, the author of this series, was one of the first writers to switch to the computer for writing. He began to use computers as early as 1988. His (relatively) early exposure to the computer and his knowledge about it made him the computer expert, and he taught many other writers to use the computer. His series is about the different experiences of different writers in learning this new medium.

To have a series published in *People's Daily* is no small feat if you are not one of the staff writers. There are, I think, three main reasons why Wu Yue's series was selected. One is that he was a famous writer. A second reason is that the topic was a hot one. A third, more important, reason is that the theme of switching writing media and the overall positive tone suited the orientation of the newspaper and of the government. Let me explain this third claim in some detail.

Of the seven articles (Wu, 1998a; Wu, 1998b; Wu, 1998c; Wu, 1998d; Wu, 1998e; Wu, 1998f; Wu, 1998g; the last one is not available) in the eight-article series, six were the success stories of about two dozen famous writers who successfully switched to computers although some of them went through some learning difficulties at first. Without exception, the use of the new medium helped these writers not only by transforming and simplifying their writing process but also by greatly increasing their productivity, no matter what difficulties they went through. One example is the story of one of the writers, Gu Jianzi, described in No. 4 of the series: "Gu Jianzi's frustrations." Gu experienced all kinds of hardware and software problems and went through three different computers until his brother bought him a multimedia PC. In addition, though he was one of the best writers in China, he was "one of the least intelligent computer learners" (Wu, 1998c). Nevertheless, after experiencing more difficulties than any other writer Wu had seen, Gu still diligently works on his writing on the computer (Wu, 1998c). For most writers, the learning process was much faster and easier, and the computer proved a great tool that improved both speed and the editing process.

One of the articles, No. 7 of the series, titled "The retreaters," does describe several failure stories. According to Wu (1998e), a superficial reason was that the computer was hard to learn and that some of the operations required the use of English. The real reason, says Wu, is that the computer does not offer the kind of freedom that the pen does. Some writers, especially poets, would like to have the pen ready so that they could write whenever their inspiration comes up. If they "had to turn on the computer, open the program, and then worry about the input method, the inspiration would already be gone" (Wu). Another writer, Cong Weixi, offered a different reason for refusing to use the computer: one of the big problems with computerization was the loss of manuscript drafts, which would prevent future readers from understanding the writer's thinking through studying the drafts. However, even this very writer later bought himself a multimedia Pentium computer and finally gave up the pen (Wu).

People's Daily is only one of the many media tools the Chinese government has available for communicating its version of knowledge about the computer and the Internet. This official interpretation of the meaning of the new technology is no doubt a dominant, though not the only, interpretation in current China. This hegemonic control of communication media has rendered the Chinese mass media a "transmission belt" (Wu, 1994, p. 210). As Wu (1994) points out:

> With the current economic reform and social change in China, the party-state will have more and more problems in its attempts to control the mass media, and the *People's Daily* will become less important than ever before with many newer, more independent sources of information. These tendencies will push the Party newspaper away from the model of command communication. At the same time, journalists will face the same restrictions in the future, because the command structure has not changed. Thus various forms of resistance will be developed. In the near future, there will be more and more possibilities for resisting political control.

What Wu describes here is still largely true today. Although the various forms of resistance he predicted are partly becoming true, especially with the popularization of the computer and Internet technology, we have reason to believe that the full realization of this prediction is still, for the large part, an optimistic dream.

10 Conclusion: Toward a More Pluralistic Model of Knowledge Construction

Behind this rapid development of the computer and Internet technology in China today, there is a hidden danger. The history of China tells us that every technological spur was followed by a long period of inertia, as Wang Yue-farn (1993) would attest. Thus, despite this seeming wonder of technological development, there is the potential threat that history may repeat itself, and this development will eventually grow into an inevitable inertia, just like previous technological developments in China. In commenting on the Qing regime, Huang (1990) has this to say:

> The regime's primary function was to hold the village communities together. Ideology came before technology, cultural influences were valued more than economics, and the passiveness of the bureaucrats as a rule took precedence over adaptability. (p. 195)

There is no reason to believe that the current Communist regime is much different from the Qing regime Huang has described here. Whether the imperial systems of the past or the Communist regime today, they all share the same structural defect: the burden of bureaucracy. As Huang (1990) points out, every time technological inertia (and social inertia in general) takes place, China is like "a submarine sandwich," with "a huge piece of bread on the top called bureaucracy" and another huge piece at the bottom "called the peasantry, both undifferentiated" (p. 193). "The makings in between, be they cultural norms or the quintessence of governance or the substance of the civil service examinations, were basically a moral platform that suited the agrarian simplicity within a country of many millions of small self cultivators" (p. 193).

Amazingly, and unfortunately, even with the social and economic progress China has enjoyed in the last twenty years or so, this sandwich structure still exists. The rapid development of computer and Internet technology, the radical enhancement of the economy, and even the considerable improvement of the welfare of the people are not the result of a deliberate social accountability on the part of the Communist government, nor a result of the inevitable tendency toward democratization; instead, they are more a by-product of the Communist effort to consolidate its hegemony. As Yue-Farn Wang (1993) points out:

> The social reward mechanism of the present system which reinforces the link between bureaucracy and prerogatives and at the same time gives low value to technical know-how naturally drives talented people away from engaging in socially relevant activities and encourages them to pursue self-interest (i.e., to become integrated once again into state power) or to pursue scientific curiosity.
>
> The immortal bureaucracy of the Chinese political system from the traditional period until the present seems to be an insurmountable monster, in spite of its early contribution to society . . . . Its destructive effect on China's development endeavors is visible. How this millennia-old tradition can be transformed in the future is, however, beyond the author's capabilities of prediction. (Y. Wang, 1993, p. 155)

Sadly, much of this is still true today. While this repetition of history on the general level does not provide us with much optimism for a democratic mode of communication in China in the immediate future, some of the unique qualities about the development of computer and Internet technology in China do seem to offer glimpses of hope for a more pluralistic model of knowledge construction in terms of the six rhetorical elements examined above. A summary of these qualities may offer us some clue as to the democratization of knowledge construction in future China.

First of all, exigency-wise, the fact that the Chinese Communist Party's political expediency to ensure its hegemony coincides with the economic and practical needs of the people is likely more than a result of mere coincidence. Obviously, it is overly simplistic and lopsided to attribute this fact exclusively either to the Communist government's political motives or to the inevitable trend toward social accountability and the force of overall development of this new technology in the world. More likely, the general trend toward democratization has forced a convergence of political motives and social accountability.

Participant-wise, though the sandwich structure that Huang (1990) described still exists, the general public today has considerably more leverage in contributing to technology development than the self-cultivating peasants in China's earlier history. The prospect of the Chinese masses playing a major role is still far from optimistic, but they are no longer an insignificant force to be easily dismissed by the bureaucrats.

Ideology-wise, what is different today is the invasion of Western ideologies that were never so influential in the earlier history of China. Though such ideologies are far from being able to become the official ideologies of the Communist government, the general public's abhorrence of the Communist ideologies and their eager, sometimes a little blind, acceptance of Western, democratic, ideologies give us every reason to believe that the pervasiveness of Western ideologies among the ordinary Chinese, and even among the Communist leaders, may be much greater than we can imagine.

In terms of knowledge construction about technology, there is also the encouraging sign that it has not been a one-party act in China. Though what computer and Internet technology means in China is largely in line with what the Communist government has mapped out, it is nevertheless also reflective of the perspectives of the ordinary users, who have been able to at least help set some of the parameters.

In terms of knowledge access and control, the power structure may be displaying certain signs of a gradual shift. Despite the fact that the political elite is largely in control of the means of power, the technical elite, though not a distinctive participant group, are achieving more power, though in a limited sense, as they move into the political elite group. As the political elite group evolves this way, it is no longer possible to say with certainty that the dynamics of this group will never change.

As for communication media, though the government still enjoys a monopoly, this monopoly is somewhat loosening up. The potential easy access to an increasing variety of media rendered possible by new technologies such as the Internet will eventually lead to more democratic use of them. In allowing the people to use the computer as a new tool of production in its effort to improve the economy, the Communist government may eventually find that it is handing over a dangerous communication weapon to the masses, the least of whose goals is to underscore the Communist hegemony.

Such a perspective may seem overly optimistic in the face of the seemingly insurmountable mountain of the Communist regime. However, the fact that the development of computer and Internet technology in China has been as much a rhetorical act as it is in any other culture, itself, suggests this development has been an act of negotiation, however small the extent of this negotiation may have been, rather than a simple act of political dictatorship. As society progresses and the democratic mechanisms slowly move into place, the seemingly insurmountable mountain of Communist hegemony can eventually be surpassed. Ironically, as Mao, the paramount dictator in China's history once said, even an ordinary, unintelligent person like Mr. Fool can remove a mountain as long as he perseveres. If history is any indication, then the 1.2 billion "fools" in China may eventually remove the mountain of political hegemony en route to a democratic, pluralistic act of knowledge construction.

Appendix: Milestone First Events in China's Internet Use

The First Chinese to Use the Internet (9/20/87): According to Xiao and Yang (2000), the first person to use the Internet in China was Qian Tianbai, who sent an email message titled "Over the Great Wall to the World" to a university in Germany on September 20, 1987. Qian is regarded as the founder of the Internet in China. This honor, however, did not come solely from the fact that he was the first to use email in China. He was also the first person to register China's top domain name CN with the Internet Center in the U.S. on behalf of China on November 28, 1990. He later also became the administrator of the CN domain. Though Qian is little known among ordinary Internet users in China, his accomplishment was by no means insignificant as his landmark actions established China's first presence on the Internet.

The First Network to Connect to the Internet (4/94): The first network in China to connect to the Internet was China's science and technology network called CNNET. The network was established in 1989, but its connection to the Internet did not occur until April 1994. A semi-governmental agency, CNNET represented the first networked organization to connect to the Internet.

The First Chinese Media to Go Online (10/20/95): The first Chinese media to go online was the newspaper *China Trade* on October 20, 1995. The interesting thing about the media in China is that many of them are either official or semi-official government agencies. Or at least, they represent the official attitude, especially on many sensitive and political issues, because their existence rests on the condition that they be compliant with the official position on those issues. Therefore, what position the government takes is first and immediately reflected in the action of the media. In this regard, the media's actions in getting online should be a good indicator of the government's policies about Internet development. However, especially interesting is the fact that *China Trade* is by no means the number one government publication in China, not even a distant second or third. Its, at most, semi-official status makes it intriguing that it is the first news media to go online, instead of the *People's Daily,* the official newspaper of the Communist Party, or CCTV, the official TV station of the government. This at least suggests that participation in Internet use at this time was not a complete monopoly by the government.

The First Internet Regulation (1/23/96): China's first Internet regulation was issued on January 23, 1996. It was titled *Temporary measures on the Internet management of information networks in the People's Republic of China.* It must be pointed out, though, that this was the first regulation on Internet use, but not the first governing information systems management. China had issued *PRC regulations on safeguarding computer information systems* as early as February 24, 1994. The function of these regulations on the Internet was more to control than to promote Internet use. This seems to go against the Chinese government's efforts to promote the development of the Internet. However, as has been the case with other new technologies, the government's efforts to promote and control the development always seem to go side by side. This represents a contradictory mindset on the part of the government in that, on the one hand, it desires the great benefits new technologies could bring, and, on the other hand, it fears the "detrimental" effects an information technology like the Internet could bring on its political regime. This love-hate relationship between new technologies and the Chinese government has characterized technological developments in China almost throughout its entire history.

The First Distance-Education Class (1996): At the beginning of the fall semester in 1996, 20 students at Fudai University in Shanghai took the first distance-education class in China. The syllabus was put on line on Fudai's website. Interestingly, the course was on computer networks. This class was obviously more an experiment than a strategic measure. Nevertheless, considering that distance education is a relatively new phenomenon even in the U.S., the offering of this class by Fudai was a ground-breaking step.

Appendix. Milestone First Events in China's Internet Use

The First E-Café/E-Bar (11/96): In November 1996, an e-café called Shi Hua Kai opened near the west gate of the Capital Stadium in Beijing. E-café was a business concept that combined the real, traditional café with computers and Internet access. People could both enjoy coffee, drinks, and a chat with friends and surf the Internet for information or pleasure. E-cafés are now very popular in China and have become, it seems, a unique Chinese phenomenon. This popularity is due to two factors. One is that cafés replaced the traditional tea houses as a popular public place of entertainment, especially for the young generation. The other factor is that at least around the mid-1990s many people still could not afford computers, or even if they had computers, they could not afford Internet access, which at that time would have cost more than a month's salary for the average Chinese, even for just limited access. E-cafés are now becoming more than just an entertainment form; they are becoming a cultural phenomenon.

The First Online Magazine (1/97): The first Chinese magazine to go online was *Wang Shang Sheng Huo* [Life on the Internet] with the English title *Internet & Intranet*. This was a monthly magazine published by The Computer and Electronics Research and Development Center of the Ministry of Electronics. Now, though accurate statistics are not available, many popular magazines are available online. Magazines' online presence represents a quite different step from that of the newspapers and other media. As I mentioned above, many newspapers in China, especially those most influential ones, are funded by the government. In their move to the Internet, loss of potential subscribers to their hard copy and therefore of revenue is not a major concern. However, for most of the magazines, which are usually self-funded, the threat of revenue loss when they go online is a real one. The switch from a charge medium to a charge-free medium is more than economic; it requires an ideological change of mindset to realize that the Internet will be an important and may be the only medium in the future for substantial increases in visibility and readership for the magazines, which may more than make up for the temporary revenue losses in the long run.

The First TV Series to Go Online (1998): The first Chinese TV series to go online was the 43-episode Shui Huo Zhuan [Waterside Legends], which was simultaneously played on CCTV and the Internet at the beginning of 1998. It was very well received by Internet users. That CCTV, the official TV station of the government, was the first to put a TV series online came as no surprise. Even in 1998, Internet TV programming was still a rather foreign concept. Now, it has become a necessary marketing tool for many TV stations.

The First Chinese Search Engine (2/15/1998): On February 15, 1998, 34-year-old Zhang Chaoyang built the first search engine in Chinese, Sohoo (with *So* meaning "search," and *hoo* borrowed from the popular U.S. search engine Yahoo!). It has become the most popular Chinese search engine. On October 5, 1998, Zhang was named by *Time* as one of the 50 "digital heroes." This was an especially important development because, prior to the birth of this search engine, Chinese Internet users had to use English to conduct searches. For the majority of Internet users in China, who had no or limited English skills, this was a daunting task. In a way, it severely limited the visibility of many sites in Chinese. The emergence of this Chinese search engine provided a new, easier-to-use medium for many users who had had difficulty accessing many sites. It proved to be a significant step in popularizing the Internet in China.

The First Large-Scale E-Commerce (6/98): During the World Cup Soccer Tournament in June 1998, China saw its first large-scale e-commerce campaign. Major networks in China competed in providing the fastest news about the tournament to the millions of fanatic soccer fans in China. Millions of Internet users visited the sites each day. This attracted many domestic and foreign businesses to advertise on these sites, which brought the Internet companies substantial profits. For the first time, the Internet business people in China began to see how the concept of a large readership brought about by free information service could bring profits. This prompted many businesses to shift their focus from short-term profit making to long-term name building.

The First Hacker (6/16/98): The first hacker incident was detected on June 16, 1998. In a routine check, technicians at an information network in Shanghai found their network had been hacked. On July 13, suspect Yang was arrested, making it the first Internet-hacking related arrest in China. Investigations revealed that the 22-year-old Yang was a graduate student in computing mathematics in the mathematics institute of a well-known university and was nationally certified in advanced software programming. Yang hacked into eight servers of the network and more than 500 customer accounts. Yang's hacking history dated back to 1996 when he used a university website and successfully hacked a science and technology network. Later, while working in a computer company, he hacked into an information network in Shanghai. The 2,000 stolen hours of Internet use alone cost the network about two thousand U.S. dollars. Yang marked the first to be arrested for "causing damage to the computer information system."

As is the case in many other hacking cases worldwide, Yang's motive was not money related, although the hacker did steal many Internet access hours. Rather, it seems the hacker did all this to prove that he had superior technical skills and software engineering knowledge, that he had the power to control and manipulate information networks if he wanted to. In a way, he was making a statement.

The First Domain Registration in Chinese (1998): In the second half of 1998, Zhong Xi Corporation introduced domain registration in Chinese for the first time in China. The company decided to offer free registration to ten government agencies and ten businesses among the first applicants. Prior to this, all domain registrations in China were done in English. What Zhong Xi did was a ground-breaking step in popularizing the Chinese language on the Internet. It also represented a significant step in China's efforts to domesticate this new technology. Such efforts may not always have been easy, for it took about thirteen years from the invention of the first China-made computer completely in the Chinese platform to the birth of a completely Chinese search engine and of the first domain registration in Chinese.

The First Online College (11/6/98): On November 6, 1998, over 10,000 students in 14 different cities and districts in Hunan province began their first class at the first online college in China—Hunan University Multimedia Information Education College. This marked a milestone in China's education history as the first full-scale distance education effort. It also provided hope for many high-school graduates for whom traditional college was still out of reach.

The First Government Website (12/16/98): The first website by any government agency in China belonged to Beijing municipal government, which opened its "Window to the Capital" site as a venue for announcing new policies and regulations and other information and also for the citizens to communicate directly with the mayor. Beijing's plan was to put all city government agencies online by the end of September 1999. Meanwhile, the State Information Department was planning to launch a national project to "put the government online" with the goal to establish a Web presence for 80 percent of the state ministries and commissions and governments at various levels by the year 2000. It should be noted that in popularizing computer and Internet technology, the Chinese government has always stayed near or at the forefront of its development.

The First Pornography Recognition Software (4/8/1999): The first pornography recognition software was introduced on April 8, 1999. This software had the ability to recognize pornography either in English or in Chinese on the Internet, on floppy diskettes, and on CD-ROMs and to stop it from displaying on the terminal. This software was of special significance to the Chinese government as it provided at least a partial solution to one of the most feared problems that came with the popularization of the Internet—people's freedom in exposure to all kinds of information, both "good" and "bad." It also represented one of the Chinese government's continuous efforts to control the kind of information people can access. And they are hardly winning the battle.

The First Internet-Related Intellectual Property Case (4/28/99): On May 10, 1998, Chen Weihua published under the pen name Wu Fang her own article titled "Interesting facts about Maya" on her personal website *3D Sesame Street.* The newspaper *Computer Commercial Information* reprinted it on October 16, 1998 without the author's permission. Chen filed a suit against *Computer Commercial Information* for its infringement on her intellectual property, and the People's Court of the Haidian District in Beijing found the defendant guilty on April 28, 1999.

Money-wise, this was a small case as it involved a total of less than $400. However, as the first case ever on Internet-related intellectual property in China, it received a great deal of attention. Its significance lies in the fact that it was the first case to attempt to define electronic intellectual property under the circumstance that relevant laws were not yet available or complete.

The First Internet TV Station (6/1/99): China's first Internet TV station, China Hongqiao Network (www.bridge.net.cn), opened its first broadcast at 8:00 PM on June 1, 1999. The program it aired that day, A New Horizon, combined traditional TV technologies with multimedia Internet technologies to provide audio, visual, and textual information. Viewers were able to communicate synchronically with the host and the guests via the Internet.

This was no doubt an experiment, and it did not turn out to be the successful entrepreneurial endeavor the investors had hoped it would be. However, the significance of the emergence of this TV network lies in that it did not have a corresponding traditional TV station. Therefore, it meant more than a traditional TV station going online. Rather, it signifies that the Internet began to be viewed in China as the fourth medium (in addition to theater, books and newspapers, and radio and TV).

References

Asian PC sales to stay strong. (2000, May 22). *China Daily*. Retrieved August 20, 2000, from http://www.chinadaily.com.cn/cndydb/2000/05/d6-1pc.522.html.
Bagley, Robert. (1999). Shang archaeology. In M. Loewe & E. L. Shaughnessy (Eds.), *The Cambridge history of ancient China* (pp. 124–231). Cambridge, UK: Cambridge University Press.
Barker, T. T., & Kemp, F.O. (1990). Network theory: A postmodern pedagogy for the writing classroom. In C.arolyn Handa (Ed.), *Computers and community: Teaching composition in the twenty-first century* (pp. 1–27). Portsmouth, NH: Boynton/Cook.
Baum, R. (1977). Diabolus ex machina: Technological development and social change in Chinese industry. In F. J. Fleron, Jr. (Ed.), *Technology and communist culture: The socio-cultural impact of technology under socialism* (pp. 315–56). New York: Praeger Publishers.
Beeching, W. A. (1974). *Century of the typewriter.* New York: St Martin's Press.
Boltz, W. G. (1999). Language and writing. In M. Loewe & E. L. Shaughnessy (Eds.), *The Cambridge history of ancient China* (pp. 74–123). Cambridge, UK: Cambridge University Press.
Bright future ahead for Chinese IT makers. (2000, July 18). *China Daily*. Retrieved August 20, 2000, from http://www.chinadaily.com.cn/cndydb/2000/07/d4-3it.718.html.
Bu, J. & Zhang, A. (1992). The Emergence of Confucianism and the contention of other various schools of thought. In Z. Tang (Ed.), *The history of Chinese civilization, Vol 3: Qin and Han* (pp. 468–506). Shijiazhuang, Hebei: Heibei Education Press.
Bu, J. & Zhang, A. (1994). Rationalism and the Prosperous Intellectual Schools. In D. Song & X. Zhang (Eds.), *The History of Chinese Civilization, Vol 6: Liao, Song, Xia, Jin* (pp. 485–549). Shijiazhuang, Hebei: Heibei Education Press.
Burke, K. (1952). *A grammar of motives.* New York: Prentice-Hall.
Burke, K. (1953). *Counter-Statement.* Los Altos, CA: Hermes Publications.
Burke, K. (1954). *Permanence and change.* Indianapolis, IN: The Bobbs-Merrill Company.

Burke, K. (1961). *Attitudes toward history.* Boston: Bacon Press.
Burke, K. (1962). *A grammar of motives and a rhetoric of motives.* Cleveland, OH: The World Publishing Company.
Burke, K. (1966). *Language as Symbolic Action.* Berkeley: University of California Press.
Burke, K. (1969). *A grammar of motives.* Berkeley: University of California Press.
Burke, K. (1972). *Dramatism and development.* Barre, MA: Clark University Press.
Burke, K. (1989). *On symbols and society.* Chicago: The University of Chicago Press.
Burns, A. (1989). *The power of the written word: The role of literacy in the history of Western civilization.* New York: Peter Lang.
Cao, M. (2000, April 29). Travel agencies turn to Internet. *China Daily*. Retrieved August 20, 2000, from http://www.chinadaily.com.cn/cndydb/2000/04/d2-2tour.429.html.
Carter, T. F. (1925). *The invention of printing in China and its spread westward.* New York: Columbia University Press.
Carter, T. F. (1955). *The invention of printing in China and its spread westward,* (2nd ed). (Revised by L. C. Goodrich). New York: The Ronald Press Company.
Ch'en, K. K. S. (1973). *The Chinese transformation of Buddhism.* Princeton, New Jersey: Princeton University Press.
Chan, S. (1985). *Buddhism in late Ch'ing political thought.* Hong Kong: The Chinese University Press.
Chen, L. (2000, June 19). Xi'an soon to become western Internet hub. *China Daily*. Retrieved August 20, 2000, from http://www.chinadaily.com.cn/cndydb/2000/06/d3-6xian.619.html.
Chen, Zhiming. (2000, March 29). Beijing has Internet in its pocket. *China Daily*. Retrieved August 20, 2000, from http://www.chinadaily.com.cn/cndydb/2000/03/d5-1wap.329.html.
Cheung, K. (1983). Recent archaeological evidence relating to the origin of Chinese characters. In D.avid N. Keightley (Ed.), *The origins of Chinese civilization* (pp. 323–91). Berkeley, CA: University of California Press.
China Internet Network Information Center (CNNIC). (2000). Major events in the development of the Internet in China. Retrieved October 12, 2002, from http://www.cnnic.net.cn/internet.shtml.
Chinese language. (1997). *Welleslian International Information Inc.* Retrieved August 20, 2000, from http://www.welleslian.com/dragontour/china/script.html.
Chu, H. (2000, January 28). China hits brakes, but E-commerce isn't slowing down; Internet: Officials are wary of freewheeling medium, but despite a host of new rules, companies say it's business as usual. *The Los*

Angeles Times. Retrieved August 20, 2000, from http://proquest.umi.com/pqdweb.

Chuang T. (1965). *Chuang Tzu*. Taipei: China Publishing House in Taiwai.

Ci Hai [An encyclopedic dictionary of the Chinese language]. (1979). Shanghai, China: Shanghai Dictionaries Press.

Cleary, T. (1992). *The essential Confucius*. San Francisco: Harper Collins.

Common facts of the history of ancient China: Special topics. (1980). Beijing, China: China Youth Press.

Cooper, J. M. (1977). The scientific and technical revolution in Soviet theory. In F. J. Fleron, Jr. (Ed.), *Technology and communist culture: The socio-cultural impact of technology under socialism*. (pp. 146–79). New York: Praeger.

Cui, J. (2000, April 3). On-line college attracts students. *China Daily*. Retrieved August 20, 2000, from http://www.chinadaily.com.cn/cndydb/2000/04/d2-4web.403.html.

Cui, N. (2000a, May 13). Computerized teaching networks aid education in central, western areas. *China Daily* Retrieved August 20, 2000, from http://www.chinadaily.com.cn/cndydb/2000/05/d2-lremo.513.html.

Cui, N. (2000b, June 28). Future education spotlights texts, teaching, technology. *China Daily*. Retrieved August 20, 2000, from ://www.chinadaily.com.cn/cndydb/2000/06/d2-3edu.628.html.

Cui, N. (2000c, March 13). Technological improvements move westward. *China Daily*. Retrieved August 20, 2000, from http://www.chinadaily.com.cn/cndydb/2000/03/d6-4tech.313.html.

Dernberger, R. F. (1977). Economic development and modernization in contemporary China: The attempt to limit dependence on the transfer of modern industrial technology from abroad and to control its corruption of the Maoist social revolution. In F. J. Fleron, Jr. (Ed.), *Technology and communist culture: The socio-cultural impact of technology underSsocialism* (pp. 224–64). New York: Praeger Publishers.

Dian, T. (2000, January 25). High-tech industries stressed. *China Daily*. Retrieved August 20, 2000, from http://www.chinadaily.com.cn/cndydb/2000/01/d1-1jia.125.html.

Doheny-Farina, S. (1992). *Rhetoric, innovation, technology: Case studies of technical communication in technology transfers*. Cambridge, Massachusetts: The MIT Press.

Domestic brands needed. (2000, May 4). *China Daily*. Retrieved August 20, 2000, from http://www.chinadaily.com.cn/cndydb/2000/05/d4-3say1.504.html.

Du, F. (1998). Oracle inscriptions. *ChinaVista*. Retrieved August 20, 2000, from http://www.chinavista.com/experience/oracle/oracle.html.

Dunn, W. N. (1977). The social context of technology assessment in eastern Europe. In F. J. Fleron, Jr., *Technology and communist culture: The socio-*

cultural impact of technology under socialism (pp. 357–96). New York: Praeger Publishers.

Ebrey, P. B. (Ed.). (1993). *Chinese civilization: A sourcebook.* New York: The Free Press.

Eisenstein, E.lizabeth L. (1979). *The printing press as an agent of change: Communications and cultural transformations in early modern Europe.* Cambridge, UK: Cambridge University Press.

Ellul, J. (1980). *The technological system* (Joachim Neugroschel, Trans.). New York: Continuum.

Fan, C. (1992). Inventions and their uses in agriculture and handicraft industry. In Z. Tang (Ed.), *The history of Chinese Civilization, Vol 3: Qin and Han* (pp. 249–308). Shijiazhuang, Hebei: Heibei Education Press.

Feenberg, A. (1975). [Review of the book *Lenine et la Revolution Culturelle*]. *Theory and Society, 2,* 599.

Feenberg, A. (1977). Transition or convergence: Communism and the paradox of development. In F. J. Fleron , Jr. (Ed.), *Technology and communist culture: The socio-cultural impact of technology under socialism* (pp. 71–114). New York: Praeger.

Feenberg, A. (1991). *Critical theory of technology.* New York: Oxford University Press.

Fei, M. Chinese dynasties. *China the Beautiful.* Retrieved February 18, 2005, from http://www.chinapage.com/main2.html.

Feng, L. (2000, June 3). Education in digital age. *China Daily.* Retrieved August 20, 2000, from http://www.chinadaily.com.cn/cndydb/2000/06/d4–3topt.603.html.

Field, M. (1977). Technology, medicine, and "veterinarism": The Western and communist experience. In F. J. Fleron, Jr. (Ed.), *Technology and communist culture: The socio-cultural impact of technology under socialism* (pp. 437–56). New York: Praeger.

Fleron, F. J., Jr. (1977a). Afterword. In F. J. Fleron, Jr. (Ed.), *Technology and communist culture: The socio-cultural impact of technology under socialism* (pp. 457–87). New York: Praeger Publishers.

Fleron, F, J., Jr. (1977b). Introduction. In F. J. Fleron, Jr. (Ed.), *Technology and communist culture: The socio-cultural impact of technology under socialism* (pp. 1–67). New York: Praeger Publishers.

Fleron, F. J., Jr. (1977c). *Technology and communist culture: The socio-cultural impact of technology under socialism.* New York: Praeger Publishers.

Forney, M (2000a, March 13). China reverses harsh Internet rules, easing threat to trade—Ban raised fears involving privacy in communications. *Wall Street Journal.* Retrieved August 20, 2000, from http://proquest.umi.com/pqdweb.

Forney, M. (2000b, February 1). Many firms fail to meet China deadline on encryption. *Wall Street Journal.* Retrieved August 20, 2000, from http://proquest.umi.com/pqdweb.

Foss, S. K., Foss, K. A., & Trapp, Robert. (1985). *Contemporary perspectives on rhetoric.* Prospect Heights, IL: Waveland Press.

Fung, Y. (1948). *A short history of Chinese philosophy* (D. Bodde, Ed.). New York: MacMillan.

Gaines, J. M. (1991). *Contested Cculture: The image, the voice, and the law.* Chapel Hill, North Carolina: The University of North Carolina Press.

Galbraith, J. K. (1971). *The new industrial state.* New York: Mentor. 1971.

Garaudy, R. (1970). *Marxism in the twentieth century.* New York: Scribner's.

General knowledge of the history of ancient China. (1980). Beijing, China: China Youth Press.

Gernet,. (1968). *Ancient China: From the beginnings to the empire* (R. Rudorff, Trans.). Berkeley, CA: University of California Press.

Green, S. (2000, March). The pen, the sword and the networked computer. *The World Today.* Retrieved August 20, 2000, from http://proquest.umi.com/pqdweb.

Guo, A. (2000, July 28). Millions switch to Internet. *China Daily.* Retrieved August 20, 2000, from http://www.chinadaily.com.cn/cndydb/2000/07/d5-3cnn.728.html.

Guo, B. (1984). *A new history of China, vol. 2.* Shanghai, China: Shanghai Renming Press.

Gusfield, J. R. (1989). Introduction. In K. Burke, *On symbols and society* (pp. 1–49). Chicago: The University of Chicago Press.

Gvishiani, D. M. (1972). *Organisation and management: A sociological analysis of Western theories.* Moscow: Progress Publishers.

Havelock, E. A. (1986). *The muse learns to write: Reflections on orality and literacy from antiquity to the present.* New Haven, CT: Yale University Press.

He, S. (2000, June 22). New tech words stir debate. *China Daily.* Retrieved August 20, 2000, from http://www.chinadaily.com.cn/cndydb/2000/06/d9-1word.622.html.

Heidegger, M. (1977). *The question concerning technology* (W. Lovitt., Trans.). New York: Harper and Row.

Heyer, P., & Crowley, David. (1991). Introduction. In H. A. Innis (Ed., 1951, Reprinted in 1991), *The bias of communication* (pp. ix-xxvi). Toronto: University of Toronto Press.

Hoffman, E.rik P. (1977). Technology, values, and political power in the Soviet Union: Do computers matter? In F. J. Fleron, Jr. (Ed.), *Technology and communist culture: The socio-cultural impact of technology under socialism* (pp. 397–436). New York: Praeger.

Huang, Ray. (1990). *China: A macro history.* Armonk, NY: M. E. Sharpe.

Hunter, D. (1947). *Papermaking: The history and technique of an ancient craft.* New York: Dover Publications.

Huo, Y. (2000a, January 20). Competitive IT industry planned. *China Daily.* Retrieved August 20, 2000, from http://www.chinadaily.com.cn/cndydb/2000/01/d2–5.120.html.

Huo, Y. (2000b, May 29). Shanghai to support high tech. *China Daily.* Retrieved August 20, 2000, from http://www.chinadaily.com.cn/cndydb/2000/05/d3–1it.529.html.

Innis, H. A. (1951). *The bias of communication.* Toronto: University of Toronto Press.

Iritani, E. (2000, March 6). Foreign capital floods China in Internet frenzy; Web: Bidding wars, reverse migration of scholars mark arrival of cyberboom. Phenomenon poses dilemma for government. *The Los Angeles Times.* Retrieved August 20, 2000, from http://proquest.umi.com/pqdweb.

Jian, B. (1979). *An outline of Chinese history, Vol. 1-2.* Beijing, China: People's Press.

Johnson, R. R. (1998). *User-centered technology: A rhetorical theory for computers and other mundane artifacts.* Albany, New York: State University of New York Press.

Johnson, S. (2001). *Emergence: The connected lives of ants, brains, cities, and software.* New York, NY: Scribner.

Keightley, D. N. (1999). The Shang: China's first historical dynasty. In M. Loewe & E. L. Shaughnessy (Eds.), *The Cambridge history of ancient China* (pp. 232–291). Cambridge, UK: Cambridge University Press.

Kohn, L. (Ed. & Trans.). (1995). *Laughing at the Tao: Debates among Buddhists and Taoists in Medieval China.* Princeton, NJ: Princeton University Press.

Lattimore, E. (1942). *China: Yesterday and today.* St. Louis, MO: Webster Publishing Company.

Lao Tzu. (1963). *Tao Te Chin.* Baltimore: Penguin Books.

LeBlanc, P. (1993). *Writing teachers writing software: Creating our place in the electronic age.* Urbana, IL: NCTE & Computers and Composition.

Lee, R. W., III. (1977). Mass innovation and communist culture: The Soviet and Chinese cases. In F. J. Fleron, Jr. (Ed.), *Technology and communist culture: The socio-cultural impact of technology under socialism* (pp. 265–311). New York: Praeger Publishers.

Leiss, W. (1972). *The domination of nature.* New York: Braziller.

Leiss, W. (1977). Technology and instrumental rationality in capitalism and socialism. In F. J. Fleron, Jr. (Ed.), *Technology and communist culture: The socio-cultural impact of technology under socialism* (pp. 115–45). New York: Praeger.

Levin, R. (1992). *Complexity: Life at the edge of chaos*. New York, NY: Macmillan.
Li, F., & Li, S. (1998, January 2). Young woman uses computer to raise chickens and makes ¥20, 000 a year. *People's Daily, The Overseas Edition*, p. 2.
Li, N. (1996, November 4–10). Moving toward the age of information: The rapid development of China's PC sector. *Beijing Review, 39*, 10–13.
Liang, C. (2000, July 26). Supercomputer holds high hopes. *China Daily* Retrieved August 20, 2000, from http://www.chinadaily.com.cn/cndydb/2000/07/d2-3comp.726.html.
Lin, Y. (Ed. & Trans.). (1938). *The Wisdom of Confucius*. New York: Random House.
Lin Yutang's invention of the Chinese typewriter. Retrieved May 16, 2003, from www.zqsunshine.com/sp45.htm.
Lin Yu-Tang's Special Invention Showroom. Retrieved February 14, 2007,. *The Lin Yu-Tang House* [On-line]. Available: http://www.linyutang.org.tw/user-en/aboutliving_1.asp.
Liu, F. (1987). The growing influence of microelectronics and computer technology in China. *Impact of Science on Society, 146*, 189–192.
Loewe, M. (1999). The heritage left to the empires. In M. Loewe & E. L. Shaughnessy, (Eds.), *The Cambridge history of ancient China* (pp. 967–1032). Cambridge, UK: Cambridge University Press.
Lu, J.. (2000, March 20). Does the Internet always mean money? *China Daily*. Retrieved August 20, 2000, from http://www.chinadaily.com.cn/cndydb/2000/03/d4-3net.320.html.
Lu, T. Ed. (1992). *A history of the Chinese civilization, Vol. 2: Pre-Qin*. Shijiazhuang, Hebei, China: Hebei Education Press.
Mao, N. (2000, June 13). Cultural feedback in China. *China Daily*. Retrieved August 20, 2000, from http:www.chinadaily.com.cn/cndydb/2000/06/d9-1feed.613.html.
McLuhan, M., & McLuhan, E. (1988). *Laws of media: The new science*. Toronto: University of Toronto Press.
Ohmann, Richard. (1985). Literacy, technology, and monopoly capital. *College English, 47*(7), 675–89.
Oliver, R. T. (1971). *Communication and culture in ancient India and China*. Syracuse: Syracuse University Press.
Ong, W. J., S.J. (1967). *The presence of the word: Some prolegomena for cultural and religious history*. Minneapolis: University of Minnesota Press.
Ong, W. J., S.J. (1971). *Rhetoric, romance, and technology: Studies in the interaction of expression and culture*. Ithaca, NY: Cornell University Press.
Ong, W. J., S.J. (1977). *Interfaces of the word: Studies in the evolution of consciousness and culture*. Ithaca, NY: Cornell University Press.

Ong, W. J., S.J. (1982). *Orality and literacy: The technologizing of the word.* London: Methuen.

Pachow, W. (1980). *Chinese Buddhism: Aspects of interaction and reinterpretation.* Laham, Maryland: University Press of America.

Peng, L. (1992). The preface. In Tao Lu (Ed.), *The history of Chinese civilization, Vol 2: Pre-Qin* (pp. 80–132). Shijiazhuang, Hebei: Heibei Education Press.

Peng, L., et al. (1992). The evolution and maturation of the Chinese language. In Tao Lu (Ed.), *The history of Chinese civilization, Vol 2: Pre-Qin* (pp. 309–367). Shijiazhuang, Hebei: Heibei Education Press.

Peng, L., et al. (Eds). (1989). *A history of the Chinese civilization, Vol. 1: Prehistory.* Shijiazhuang, Hebei, China: Hebei Education Press.

Platt, K. (2000, June 1). With a click, Chinese vault cultural walls. *Christian Science Monitor.* Retrieved August 20, 2000, from http://proquest.umi.com/pqdweb.

Pomfret, J. (2000, February 13). Chinese web opens portals to new way of life; Booming Internet splits haves and have-nots. *The Washington Post.* Retrieved August 20, 2000, from http://proquest.umi.com/pqdweb.

PRC Yearbook '98/99. (1998/99). Beijing, China: PRC Yearbook Ltd.

Pre-Qin calligraphy: Foundation for Chinese calligraphy. *XinYuSi* Retrieved October 13, 2006 from http://www.xys.org/xys/ebooks/others/history/calligraphy_history/shufa01.txt.

Qiao, W. (1994). The adjustment and maturation of the feudal educational system. In D. Song & X. Zhang (Eds.), *The history of Chinese civilization, Vol 6: Liao, Song, Xia, and Jin Dynasties* (pp. 440–484). Shijiazhuang, Hebei: Heibei Education Press.

Qiao, W., & Liu, H. (1992). The rise and fall of elite schools and the popularization of private schools. In T. Lu (Ed.), *The history of Chinese civilization, Vol 2: Pre-Qin* (pp. 368–391). Shijiazhuang, Hebei: Heibei Education Press.

Qin, J. (1997, January 25). Global Chinese computer information network established. *People's Daily, The Overseas Edition*, p. 4.

Quadrupled in a year: The number of netizens in China reaches 8.9 million. (2000, January 20). *Sina News Center.* Retrieved August 20, 2000, from http://dailynews.sina.com/china/kwongzhou/2000/0120/665918.html.

Rawson, J. (1999). Western Zhou archaeology. In M. Loewe & E. L. Shaughnessy (Eds.), *The Cambridge history of ancient China* (pp. 352–449). Cambridge, UK: Cambridge University Press.

Ren, T. (1997, March 7).' 97 international computer fair (China) the largest ever. *People's Daily, The Overseas Edition*, p. 4.

Richta, R. (1969). *Civilization at the crossroads; Social and human implications of the scientific and technological revolution* (M. Slingova, Trans.). White Plains, NY: International Arts and Sciences Press.

Rodzinski, W. (1984). *The walled kingdom: A history of China from antiquity to the present.* New York: The Free Press.

Selfe, C. L., & Selfe, R. J., Jr. (1994). The politics of the interface: Power and its exercise in electronic contact zones. *College Composition and Communication, 45*(4), 480–504.

Shang, X.. (1992). The rich and varied literature and art. In T. Lu (Ed.) *The history of Chinese civilization, Vol 2: Pre-Qin* (pp. 473–503). Shijiazhuang, Hebei: Heibei Education Press.

Shaughnessy, E.dward L. (1999a). Calendar and chronology. In M. Loewe & E. L. Shaughnessy, (Eds.), *The Cambridge history of ancient China* (pp. 19–29). Cambridge, UK: Cambridge University Press.

Shaughnessy, E.L. (1999b). Western Zhou history. In M. Loewe & E. L. Shaughnessy, (Eds.), *The Cambridge history of ancient China* (pp. 292–351). Cambridge, UK: Cambridge University Press.

Shen, C. (1989). The origin of Chinese language. (In L. Peng, J. Qi, & C. Fan (Eds.), *The history of Chinese civilization, Vol 1: Prehistory* (pp. 427–437). Shijiazhuang, Hebei: Heibei Education Press.

Short-term goals of the Chinese software industry. (1997, February 10). *People's Daily, The Overseas Edition,* p. 2.

Slack, J. D. (1984). *Communication technologies and society: conceptions of causality and the politics of technological intervention.* Norwood, NJ: Ablex Publishing Corporation.

Smith, C. S. (2000, June 7). Three roads in China: B2B, B2C and C2C. *New York Times.* Retrieved August 20, 2000, from http://proquest.umi.com/pqdweb.

Somers, R. M. (1990). Introduction: Buddhism and Chinese culture. In A. F. Wright (Ed.), *Studies in Chinese Buddhism.* New Haven, CT: Yale University Press.

Stocks cheered up by officials' rosy remarks. (2000, April 20). *China Daily.* Retrieved August 21, 2000, from http://www.chinadaily.com.cn/cndydb/2000/04/d7-1stok.420.html.

Suchman, L. A. (1987). *Plans and situated actions: The problem of human-machine communication.* Cambridge: Cambridge University Press.

Sui, C. (2000, February 1). Chinese rule confuses firms; Effort to control use of encryption widely ignored. *The Washington Post.* Retrieved August 21, 2000, from http://proquest.umi.com/pqdweb.

Sullivan, P. & Porter, J. E. (1997). *Opening spaces: Writing technologies and critical research practices.* Greenwich, CT: Ablex.

Sun, M. (2000, January 30). Cruising the net for jobs. *China Daily.* Retrieved August 21, 2000, from http://www.chinadaily.com.cn/cndydb/2000/01/b2-2hunt.130.html.

Sun, S. (2000, January 19). Young, smart, single and surfing. *China Daily.* Retrieved August 21, 2000, from http://www.chinadaily.com.cn/cndy-db/2000/01/d3-2net.119.html.

Survey of Internet development in China, No. 1. (1997, October). *China Internet Network Information Center* [On-line]. Available: http://www.cnnic.org.cn/download/2003/10/13/93603.pdf.

Survey of Internet development in China, No. 2. (1998, July). *China Internet Network Information Center* [On-line]. Available: http://www.cnnic.org.cn/download/2003/10/13/92926.pdf.

Survey of Internet development in China, No. 3. (1999, January). *China Internet Network Information Center* [On-line]. Available: http://www.cnnic.org.cn/download/2003/10/13/93056.pdf.

Survey of Internet development in China, No. 4. (1999, July). *China Internet Network Information Center* [On-line]. Available: http://www.cnnic.org.cn/download/2003/10/13/92809.pdf.

Survey of Internet development in China, No. 5. (2000, January). *China Internet Network Information Center* [On-line]. Available: http://www.cnnic.org.cn/download/2003/10/13/92638.pdf.

Survey of Internet development in China, No. 6. (2000, July). *China Internet Network Information Center* [On-line]. Available: http://www.cnnic.org.cn/download/2003/10/13/91748.pdf.

Survey of Internet development in China, No. 7. (2001, January). *China Internet Network Information Center* [On-line]. Available: http://www.cnnic.org.cn/download/2003/10/13/91443.pdf.

Survey of Internet development in China, No. 8. (2001, July). *China Internet Network Information Center* [On-line]. Available: http://www.cnnic.org.cn/download/2003/10/10/171539.pdf.

Survey of Internet development in China, No. 9. (2002, January). *China Internet Network Information Center* [On-line]. Available: http://www.cnnic.org.cn/download/2003/10/10/171802.pdf.

Survey of Internet development in China, No. 10. (2002, July). *China Internet Network Information Center* [On-line]. Available: http://www.cnnic.org.cn/download/2003/10/10/171730.pdf.

Survey of Internet development in China, No. 11. (2003, January). *China Internet Network Information Center* [On-line]. Available: http://www.cnnic.org.cn/download/2003/10/10/170932.pdf.

Survey of Internet development in China, No. 12. (2003, July). *China Internet Network Information Center* [On-line]. Available: http://www.cnnic.org.cn/download/2003/10/10/171934.pdf.

Survey of Internet development in China, No. 13. (2004, January). *China Internet Network Information Center* [On-line]. Available: http://www.cnnic.org.cn/download/manual/statisticalreport13th.pdf.

References

Survey of Internet development in China, No. 14. (2004, July). *China Internet Network Information Center* [On-line]. Available: http://www.cnnic.org.cn/download/2004/2004072002.pdf.

Survey of Internet development in China, No. 15. (2005, January). *China Internet Network Information Center* [On-line]. Available: http://www.cnnic.org.cn/download/2005/2005011801.pdf.

Survey of Internet development in China, No. 16. (2005, July). *China Internet Network Information Center* [On-line]. Available: http://www.cnnic.org.cn/uploadfiles/pdf/2005/7/20/210342.pdf.

Survey of Internet development in China, No. 17. (2006, January). *China Internet Network Information Center* [On-line]. Available: http://www.cnnic.org.cn/images/2006/download/2006011701.pdf

Survey of Internet development in China, No. 18. (2006, July). *China Internet Network Information Center* [On-line]. Available: http://www.cnnic.org.cn/uploadfiles/pdf/2006/7/19/103651.pdf.

Survey of Internet development in China, No. 19. (2007, January). *China Internet Network Information Center.* Retrieved February 5, 2009, from http://www.cnnic.net.cn/uploadfiles/doc/2007/1/23/113530.doc.

Technology plan set for 1996–2000. (1995, November 27-December 3). *Beijing Review, 38*(48), 4–5.

Tsien, T.. (1962). *Written on bamboo and silk: The beginning of Chinese books and inscriptions.* Chicago: University of Chicago Press.

Tsien, T. (2004). *Written on bamboo and silk: The beginning of Chinese books and inscriptions, 2nd Edition.* Chicago: University of Chicago Press.

Twitchett, D. (1983). *Printing and publishing in medieval China.* New York: Frederic C. Beil.

Van Maanen, J. (1988). *Tales of the field: On writing ethnography.* Chicago: The University of Chicago Press.

Vista changes the PC market. (2007, February 26). Retrieved February 5, 2009, from http://discovery.ynet.com/view.jsp?oid=19009682.

Wang, H. (1993). *Chinese public discourse: A rhetorical analysis of the account of the Tiananmen Square incident by the newspaper People's Daily.* Unpublished doctoral dissertation, Purdue University, West Lafayette, Indiana.

Wang, N., Li, G .& Zhu, X. (1992). In-depth studies in language and the emergence of book systematics. In T. Zangong, (Ed.), *The history of Chinese civilization, Vol 3: Qin and Han* (pp. 412–440). Shijiazhuang, Hebei: Heibei Education Press.

Wang, Y. (1993). *China's science and technology policy: 1949–1989.* Aldershot, England: Avebury.

Wang, Y. & Cheng, G. (2000, April). E-commerce: China brewing a new economic revolution. *China News Digest, 471.* Retrieved February 5, 2009, from http://www.cnd.org/HXWZ/new.hz8.htm.

Wang, Z. Lin Yutang and Pearl Buck. Retrieved May 16, 2003, from http://www.chinawriter.org/zjljl/hwzj/hwzj000020.htm.

Ware, J. R. (1955). Introduction. In J. R. Ware (Trans.), *The Sayings of Confucius*. New York: New American Library.

Williams, F. & Gibson, D. V. (1990). *Technology transfer: A communication perspective*. Newbury Park, CA: Sage.

Williams, R. (1983). *Keywords: A vocabulary of culture and society, Revised edition*. New York: Oxford University Press.

Winner, L. (1977). *Autonomous technology: Technics-out-of-control as a theme in political thought*. Cambridge, MA: The MIT Press.

Winner, L. (1986). *The whale and the reactor: A search for limits in an age of high technology*. Chicago: The University of Chicago Press.

Winograd, T, & Flores, F. (1986). *Understanding computers and cognition: A new foundation for design*. Norwood, NJ: Ablex.

Wright, A. F. (1990). *Studies in Chinese Buddhism*. New Haven, Connecticut: Yale University Press.

Wu, G. (1994, March). Command communication: The politics of editorial formulation in the *People's Daily*. *China Quarterly, 137,* 194–211.

Wu, Y. (1998a, August 8). The switching of "pens" by Chinese writers (5): The best among the "pen" switchers. *People's Daily, The Overseas Edition*, p. 2.

Wu, Y. (1998b, April 18). The switching of "pens" by Chinese writers (1): From the very beginning. *People's Daily, The Overseas Edition*, p. 2.

Wu, Y. (1998c, July 4). The switching of "pens" by Chinese writers (4): Gu Jian Zi's frustrations. *People's Daily, The Overseas Edition*, p. 2.

Wu, Y. (1998d, May 9). The switching of "pens" by Chinese writers (2): Like a tiger with wings. *People's Daily, The Overseas Edition*, p. 2.

Wu, Y. (1998e, September 15). The switching of "pens" by Chinese writers (7): The retreaters. *People's Daily, The Overseas Edition*, p. 2.

Wu, Y. (1998f, August 22). The switching of "pens" by Chinese writers (6): The unforgettable students. *People's Daily, The Overseas Edition*, p. 2.

Wu, Y. (1998g, June 6). The switching of "pens" by Chinese writers (3): Where there is a will there is a way. *People's Daily, The Overseas Edition*, p. 2.

Xia, Z., et al. (1979). *Ci Hai, Suo Yin Ben* [*Sea of Words, The compressed edition*]. Shanghai, China: Shanghai Lexicography Press.

Xiao, J. (2000, June 19). Cyber plays a role in sex education. *China Daily*. Retrieved August 21, 2000, from http://www.chinadaily.com.cn/cndy-db/2000/06/d9-2sex.619.html.

Xiao, P. & Yang, Y. (2000). The first's of the Internet in China. *Mei Zhou Wen Hui Zhou Kan* [Sinotimes]. Retrieved August 21, 2000, from http://www.sinotimes.com/100/142.htm#5.

Xie, S., & Zhang, D. (2000, May 11). Premier vows to speed up opening. *China Daily.* Retrieved August 21, 2000, from http://www.chinadaily.com.cn/cndydb/2000/05/d1-1zhu.511.html.

Xu, L. (2000, March 17). Reins on e-cafes tightened. *China Daily.* Retrieved August 21, 2000, from http://www.chinadaily.com.cn/cndydb/2000/03/d3-1int.317.html.

Xu, X. (2000, May 19). B2B, B2C, Shanghai to E. *China Daily.* Retrieved August 21, 2000, from http://www.chinadaily.com.cn/cndydb/2000/05/d6-2sh.519.html.

Xue, H. (2000, January 8). E-commerce needs legal support. *China Daily.* Retrieved August 21, 2000, from http://www.chinadaily.com.cn/cndydb/2000/01/d4-1eco.108.html.

Yan, B. (1997, January 23). China strengthens further development for its information industry: Electronic market capacity to exceed ¥ 1000 billion. *People's Daily, The Overseas Edition,* p. 1.

Yang, M. (2005). New breakthroughs in the research of oracle inscriptions. *China News Digest.* Retrieved August 21, 2000, from http://my.cnd.org/modules/wfsection/article.php?articleid=10489.

Yates, J. (1989). *Control through communication: The rise of system in American management.* Baltimore, Maryland: The Johns Hopkins University Press.

Yu, Q. (1992). Unified educational policy and system. (T. Zangong, Ed.), *The history of Chinese civilization, Vol 3: Qin and Han* (pp. 391–411). Shijiazhuang, Hebei: Heibei Education Press.

Zeng, M. (2000, April 15). Tech-aided farming coming. *China Daily.* Retrieved August 21, 2000, from http://www.chinadaily.com.cn/cndydb/2000/04/d2-1sh.415.html.

Zhang, J. X., & Wang, Y. (1995). *The emerging market of China's computer industry.* Westport, CT: Quorum Books.

Zhang, Y. (2000, March 17). Computer code rule decoded. *China Daily.* Retrieved August 21, 2000, from http://www.chinadaily.com.cn/cndydb/2000/03/d1-code.317.html.

Zhong, S. (1992). Rapid development in the military and the rising military science. In T. Lu (Ed.). *A history of the Chinese civilization, Vol. 2: Pr- Qin* (pp. 80–132). Shijiazhuang, Hebei, China: Hebei Education Press.

Zhu, C., & Tan, H. (2000, March 24). Port city boosts hi-tech. *China Daily.* Retrieved August 21, 2000, from http://www.chinadaily.com.cn/cndydb/2000/03/d5-6dd.324.html.

Zhu, Q. (2000, May 8). Culture industry needs protection. *China Daily.* Retrieved August 21, 2000, from http://www.chinadaily.com.cn/cndydb/2000/05/d4-1cul.508.html.

Index

act, 7, 29, 34, 40, 55, 60, 62-63, 69-80, 102, 107, 115, 117, 131-132, 217, 233, 234
action, 61, 69
aesthetic investment, 57, 58
agency, 70, 71, 77, 79, 80, 164, 226, 235, 240
agent, 38, 70, 71, 75, 76, 77, 78, 79, 80, 87, 160, 198, 207, 208
Althusser, Louis, 84, 85
ambivalence theory, 39, 47- 50, 52-53, 59, 218
animality, 69, 77
Anyang, 106, 114
Asia Minor, 120
associative compounds, 9, 108, 111
attitude, 16, 39, 78-80, 87, 101, 148, 156, 193, 194, 220, 235
automation, 52-53

Bagley, Robert, 119
Bai Ju Yi, 156
ballpoint pens, 32
bamboo, 4, 5, 119, 120-121, 123-129, 133-134, 136, 138-140, 154
ban (版), 125
ban du (版牍), 125
Baum, Richard, 28, 42-43, 98
Beijing (Peking) Man, 112
Beijing Applied Computer Research Institute, 202
Beijing University, 202

Bi Sheng, 154-155, 158, 159, 160, 164
block printing, 5, 105, 151-154, 155, 156, 157, 158, 160, 165
Boltz, William G., 8, 119
book taxonomy, 150
bronze inscriptions, 4, 103, 104, 109-112, 114-119, 128, 129, 130, 132, 133; uses, 109-111
brush pen, 5, 32, 118-123, 125, 130, 142, 153
Bu, Jinzhi & Zhang, Anqi, 115, 130
Buddhism, 10, 19, 20- 23, 114, 145, 147, 152, 163, 192, 193; arguments against Taoism, 21-22
Buddhist rhetoric, 23, 193; view of language, 23
Burke, Kenneth, 67-80
Burke's pentad: act, 7, 29, 34, 40, 55, 60, 62-63, 69-80, 102, 107, 115, 117, 131, 132, 217, 233, 234; agency, 70-71, 77, 79 80, 164, 226, 235, 240; agent, 38, 70, 71, 75, 76, 77, 78-79, 80, 87, 160, 198, 207, 208; scene, 5, 22-23, 27, 67, 70-72, 74, 75, 79-80, 183, 190
Burns, Alfred, 102

Cai Lun, 124, 137, 138, 139, 140, 147

257

calligraphy, 32, 141, 142, 144, 156, 158, 161, 162
Cangjie, 8
CCTV, 236, 237
ce (策), 125
Chan, Sin-wai, 20
Changsha, 120, 121, 138
Chen Wei, 143
Cheung, Kwong-Yue, 8
China Education and Research Network (CERNET), 205, 206, 207
China Golden Bridge Network (CHINAGBN), 206
China Information Safety Testing and Examination Center (CNISTEC), 207
China Internet Network Information Center (CNNIC), 202, 206
China National Committee for Terms in Science and Technologies (CNCTST), 223
China Research Network (CRN), 202
China Securities Regulatory Commission, 207
China Telecommunications, 204, 205
CHINANET, 205, 207
Chinese Academic Network (CANET), 202
Chinese characters, 7, 9, 30, 105, 111, 120, 162, 167, 169, 174, 176, 180; formation, 9; input method, 170, 176, 178, 180, 230; style, 149
Chinese government, 76, 194, 198, 199, 200, 201, 202, 207, 208, 209, 210, 211, 215, 218, 221, 223, 230, 236, 240
Chinese script, 7, 8, 9, 108, 144, 150, 161, 167, 176; origin, 8;

six features, 108
Chinese typewriter, 4, 105, 167, 168, 169, 170, 171, 172, 173, 174, 175, 176, 192; character arrangement, 169, 173, 174, 175; character plate, 172; ink ribbon, 172; typing key, 172, 173
Chinese typography, 162
Chu poetry, 135
Chu shu chi nien, 121
Chuang Tzu, 15, 16, 17
Chuang Tzu, 121
class distinction, 94, 95
Cleary, Thomas, 11, 12, 13
cognitive understanding, 41, 47
coin inscriptions, 128
collegiality, 57, 58
Common Facts of the History of Ancient China: Special Topics, 106, 123, 125, 126, 152
communication medium, 62, 77, 99-103, 116, 117, 134-135, 149-150, 165-166, 186, 224-230, 241
Communist technical code, 217, 218
Computer and Electronics Research and Development Center of the Ministry of Electronics, 237
Computer Commercial Information, 240
computer hacking, 219, 238, 239
Computer Network Information Center, 183, 203, 204
computer sales, 180, 181, 183
concretization, 51, 57, 58, 195
Confucian classics, 143, 145, 146, 147, 148, 157, 161, 164
Confucian rhetoric, 14, 193
Confucianism, 10, 11, 13, 14, 15, 17, 18, 20, 21, 22, 23, 114,

Index   259

130, 131, 135, 143, 144, 145, 146, 147, 148, 157, 161, 163, 164, 192, 193
Confucius, 10, 11, 12, 13, 133, 134, 164; social order and harmony, 11
Cong, Weixi, 230
convergence theory, 39, 41, 42, 43, 48; propositions, 43
Cooper, Julian M., 40, 44, 45, 46
critical theory of technology, 27, 39, 53, 54, 59; differences from neutrality theory and technological determinism, 59; instrumentalization, 54, 55, 57, 58, 59
CSTNET, 206, 207
cultivation of the individual, 11, 12
cultural context, 4, 6, 30, 35, 36, 54
cultural ethos, 90, 91, 92, 98
cultural knowledge, 91
Cultural Revolution, 187, 189
culture: definition, 33-34, 91; relationship with knowledge, 91-92
cybernetics, 52, 53

Daedalus, 35, 36
Dawenkou Culture, 5, 8, 106
DaYuDing, 110
DECnet, 202
decontextualization, 31, 54, 55, 56, 57, 59
Deng Xiaoping, 143, 187, 189, 190, 195, 215, 216
Deng Yu, 143
Dernberger, R. F., 28, 42, 92
dharani, 152
Dialects, 150
Dian, Tai, 207
Dingcun Man, 112

distribution of knowledge, 93, 99-102, 224-230
divination, 106, 107, 108, 112, 113, 115, 116, 132, 143
Doheny-Farina, Stephen, 29, 35, 36, 66, 88, 89; *Rhetoric, Innovation, Technology; Case Studies of Technical Communication in Technology Transfers*, 66
domain, 185, 203, 204, 206, 211, 227, 235, 239
Dong Zhong Shu, 145
Dongfang Su, 126
dramatism, 69, 70, 73
du (牍), 125
Du Fu, 156
Du Mu, 156
Du Pont, 86, 87
Du, Feibo, 9, 106, 107, 108
Dunhuang, 137, 152
Dunn, William, 28, 35, 91

Eastern Han Dynasty, 139, 140, 141, 143, 147
Eastern Zhou Dynasty, 118
Ebrey, Patricia B., 17, 19, 20
education, 11, 69, 101, 113, 114, 129, 132, 133, 134, 135, 136, 143, 144, 146, 147, 148, 149, 159, 161, 164, 191, 208, 239
Egypt, 103, 120, 140, 151, 165
Eisenstein, Elizabeth, 101, 102
Ellul, Jaques, 42, 59
Emperor Qin, 126, 128, 135, 141, 144, 148
Emperor Tang Tai Zong, 156
encultured consciousness, 3
Er Ya (Standard Chinese) 《尔雅》, 150
Ethics and Politics 《大学》, 10
etymology, 150
exigency, 74, 75, 80, 81, 82, 83, 86, 87, 102, 103, 112, 113,

129, 130, 141, 142, 144, 157, 160, 176, 186, 188, 191, 192, 197, 232
exploratory interaction, 47-53

Fan Ye, 137, 138
Feenberg, Andrew, 27, 28, 31, 35, 36, 40, 41, 43, 47, 48, 49, 53, 54, 55, 56, 57, 58, 59, 66, 86, 91, 94, 95, 98, 100, 103, 214, 217, 218; *Critical Theory of Technology*, 27, 31, 36, 39, 53, 54, 59
Feng Dao, 153, 157
Field, Mark, 42
Five Dynasties, 105, 164
Fleron, Frederic J., Jr, 26, 28, 35, 39, 40, 41, 50, 51, 52, 53, 91, 93, 99
foreign transfer, 4, 6, 176, 178
Forney, Matt, 211, 216, 217
Foss, Sonja K., Foss, Karen A., & Trapp, Robert, 33, 71
fountain pens, 32
Fung, Yulan, 19, 20

Galbraith, John K., 42
Gansu, 137
Gao, Xiqing, 207
Garaudy, Roger, 93
Germany Research Network (GRN), 202
Golden Bridge Project, 203, 204, 205
Green, Stephen, 196, 211
Gu, Jianzi, 229
Guan Zi 《管子》, 125
Guo Yu, 135
Guo, Bonian, 165; *A New History of China*, 152
Guo, Moruo, 106
Gutenberg, 155, 165, 166
Gvishiani, Dzherman M., 45, 46

Han Dyansty, 104, 120, 125, 137, 139, 141, 142, 143, 144, 145, 146, 147, 148, 149, 150, 163
Han Fei, 131, 135
Han Fei Zi 《韩非子》, 135
Havelock, Eric A., 102
Heidegger, Martin, 42, 59
hemp paper, 137, 138, 139, 140
Hetao Man, 112
hieroglyphics, 9, 108, 124
High Energy Physics Institute, 203, 204
Hoffman, Erik P., 35, 96, 97
homogeneity, 163
Hongqiao Network, 241
Hou Han Shu (*The History of the Late Han Dynasty*) 《后汉书》, 137, 138
Hou, Mingjuan, 183, 201, 208
Hu Qili, 200
Huang, Ray, 113, 142, 143, 162, 163, 208, 231, 233
humanity, 12
Hunter, Dard, 119, 120, 123, 126, 140, 142, 152, 162

id, 79
ideo-logic, 98
ideology, 4, 6, 10, 14, 19, 21, 23, 27, 43, 51, 59, 73, 78, 79, 80, 83, 84, 85, 86, 87, 91, 101, 102, 103, 114, 118, 130, 131, 132, 144, 145, 148, 158, 159, 160, 162, 163, 165, 186, 192, 194, 196, 199, 211, 218, 231, 233
Imperial College, 141, 142, 143, 144, 146, 148, 149, 164
imperial seal inscriptions, 128
informal information model, 88
information gatekeeping model, 88

Index

information interface model, 88
information transfer model, 88
ink, 4, 5, 32, 104, 118, 120, 121, 122, 123, 125, 133, 134, 153, 165, 172; soot, 5, 119, 121, 122
ink slab, 121, 123, 133
Innis, Harold, 61; *The Bias of Communication*, 61
instrumentalization, 54, 55, 57, 58, 59
intellectual elite, 19, 145, 146, 147, 213
intellectual property, 185, 240
intellectual studies, 145, 146, 148, 159, 161, 164
intellectuals, 131, 132, 134, 142, 143, 144, 148, 159, 168, 213, 223
interiorization, 63
Internet: laws and regulations, 210

jade and stone inscriptions, 128
Japan, 140, 152, 165, 180, 203
jen (仁humanity), 12, 193
jian (简), 125
jian ce (简策), 125
Jian, Bozhan, 8, 10, 108, 112, 113, 114, 115, 131, 140, 141, 142, 156
Jiang Zemin, 207
Johnson, Robert R., 64, 65, 66
Johnson, Steven, 37, 38
Joint Commission for China-U.S. Science and Technology Cooperation, 204

Karma, 19, 21
Keep Commission, 83
Keightley, David N., 108
knowledge access and control, 49, 77-78, 80, 81, 92-99, 102, 103, 116, 133-134, 148-149, 164-165, 186, 210, 211, 220, 221-224, 225, 233, 240
knowledge creation, 4, 35, 36, 60, 61, 62, 66, 77, 78, 80, 81, 89- 92, 93, 95, 96, 97, 98, 99, 100, 101, 102, 103, 115-116, 132-133, 147, 161-164, 186, 217- 221, 222, 223, 224, 231, 232, 233, 234
Kohn, Livia, 21, 22
Korea, 140, 165

Land Regulations, 126
language: ideogram-based, 7
language, phoneme-based, 7
Lao Tzu, 15, 17, 131
Lao Tzu 《老子》, 135
Later Han Dynasty, 143
Lattimore, Eleanor, 8, 103, 107, 108, 118, 160
LeBlanc, Paul, 35, 36
Lee, Rensselaer, 28, 42, 213
Legalism, 10, 131
Leiss, William, 24, 25, 26, 28, 35
Leninism, 10
Levin, Roger, 37
li (礼Letiquette), 12, 13, 193
Li Bai, 156
Li Peng, 203
Li Sao, 135
Li Shang Yin, 156
Li, Fengyan & Li, Shijie, 228
Li, Ning, 180, 181, 201
Liangzhu Culture, 106
Lin, Yutang, 11, 12, 13, 168, 169, 170
Liu Xiu, 143
Liu, Fengqiao, 179, 180, 190, 191
Loewe, Michael, 130
Longshan Culture, 106, 112
Lu, Jingxian, 177, 208

Majiabang Culture, 106
Majiayao Culture, 106

Mao, Ning, 220
Mao, Zedong, 91, 187, 189, 190, 215, 234
Maoism, 10, 23, 189, 192, 195
Marx, 48, 49, 94, 95
Marxism, 10, 48, 192, 194, 195, 196, 213, 222
McLuhan, Marshall & McLuhan, Eric, 62; *Laws of Media*; *The New Science*, 62
mediation theory, 39, 47, 50, 51, 52, 53
medium, 62, 77, 100, 101, 102, 113, 114, 116, 117, 126, 135, 140, 147, 150, 156, 165, 168, 186, 224, 228, 229, 237, 238, 241
Mencius, 11, 121, 131, 134
Meng Hao Ran, 156
Meng Tien, 119
Meng Xi Bi Tan (Dream Pool Jottings), 160
Ming Dynasty, 105, 158
Ming Kuai Typewriter (The Chinese Fast Typewriter), 170
Mo Zi, 131
Mo Zi 《墨子》, 135
movable type, 4, 5, 151-154, 155, 157, 158, 159, 160, 161, 162, 164, 165, 167
mutually explanatory or synonymous characters, 108

National Informationalization Plan, 206
native development, 4, 6, 35, 178, 181
NCFC, 202, 204
Neo-Confucianism, 144, 145, 163
Neolithic Culture, 112
Neolithic Period, 112
neutrality theory, 39- 41, 44, 48, 49, 50, 54, 59; dismissal of technological impact, 40, 41

Nirvana, 20
Northern and Southern Dynasty, 137

Ohmann, Richard, 60
Oliver, Robert T, 3, 11, 13
Ong, Walter, 62, 102
ontological design, 61
oracle inscription, 4, 5, 8, 9, 103-109, 111, 112, 113, 114, 116, 117, 129, 130, 132, 160

Pachow, W, 19
Paleolithic Period, 112
paper, 4, 5, 104, 118, 119, 130, 136-150, 153, 156, 158, 160, 165; bamboo, 4, 5, 119, 120, 121, 122, 123-126, 127, 128, 129, 133, 134, 136, 138, 139, 140, 154; early forms, 123-129; origin, 137-138; roll, 5, 120; silk, 4, 5, 120, 122, 123, 124, 126-127, 128, 129, 130, 133, 134, 136, 138, 140, 143, 151; wood slips, 4, 5, 119, 120, 123-126, 128, 129, 152
papermaking process, 139-140
papyrus, 120, 124
parchment, 120, 124
participant, 4, 58, 60, 64, 65, 66, 75, 76, 78, 80, 84, 85, 87, 88, 89, 90, 92, 93, 96, 99, 100, 102, 103, 115, 116, 132, 145, 146, 178, 186, 192, 197, 198, 212, 214, 217, 220, 221, 222, 224
passive exposure, 38, 39
Peacocks Flying to Southeast 《孔雀东南飞》, 147
pen, 4, 5, 32, 118, 119-120, 123, 130, 133, 134, 168, 186, 228, 229, 230, 240; bamboo, 5, 119, 120

Index

Peng, Lin, et al, 5, 107, 109, 110, 111, 112, 118, 125, 126, 129
pentad, 67, 68, 70, 74, 75, 78, 79, 80
phonetic loan characters, 108
pictophonetics, 9, 108
Platt, Kevin, 196, 219
political elite, 78, 90, 93, 95, 98, 99, 132, 197, 198, 212, 213, 217, 221, 222, 233
political ideology, 94, 145, 159, 189, 192, 194, 195, 196
Political Strategists, 144
Pomfret, John, 184, 196, 219
pornography, 185, 219, 240
positioning, 55, 56, 96
pottery, 8, 105, 119, 121, 123, 128, 138
pottery inscription, 128
press copying, 82, 83, 86
primary instrumentalization, 54, 55, 58, 59; automation, 52, 53; decontextualization, 31, 54, 55, 56, 57, 59; positioning, 55, 56, 96; reductionism, 55, 56, 58
printing, 4, 5, 6, 77, 101, 105, 149, 151-166, 167, 172
private school, 132, 133, 134, 135, 144, 148, 149
proactive control, 38, 53, 67
Prose, 35, 36
Public Multimedia Communication Network (169 Net), 206

Qian, Hualin, 203
Qian, Tianbai, 202, 203, 204, 235
Qiao, Weiping & Liu, Hong, 132, 133, 134
Qimin Yaoshu, 122
Qin Dynasty, 104, 119, 120, 123, 125, 126, 127, 128, 129, 133, 135, 139, 141, 144, 148, 149, 152

Qin Shihuang, 126, 128
Qin, Jingwu, 184
Qinghua University, 202
Qu Yuan, 135
quill pen, 86

ratio, 71, 72, 74, 75, 76, 77, 78
Rawson, Jessica, 120
reductionism, 55, 56, 58
Remington, 169
rhetorical perspectives on technology, 23, 36, 39, 53, 60, 62, 64, 66, 102
Richta, Rodavan, et al., 99
Rodzinski, Witold, 8, 107
role of experts, 96, 97
roll, 5, 120
royal decrees, 128

Samsara, 20
scene, 5, 67, 70, 71, 72, 74, 75, 79, 80
secondary instrumentalization: aesthetic investment, 57, 58; collegiality, 57, 58; concretization, 51, 57, 58, 195; vocation, 57, 58
self assertion, 158, 161, 163
Selfe, Cynthia & Selfe, Richard, 63
self-explanatory characters, 108
Shang Dynasty, 5, 9, 104, 106, 107, 108, 109, 110, 111, 112, 113, 114, 115, 116, 117, 118, 119, 120, 121, 125, 128, 129, 130, 131, 132, 137
Shang, Xuefeng, 135
Shaughnessy, Edward L., 110, 124
Shen Kuo, 155, 160
Shen, Changyun, 105, 106
Shi Ming (*Interpretation of Names*), 150
Shiji (*Records of the Grand Historian*) 《史记》, 154

Shuo Wen Jie Zi (*Annotated Dictionary of the Chinese Language*) 《说文解字》, 150
silk, 4, 5, 120, 122, 123, 124, 126, 127, 128, 129, 130, 133, 134, 136, 138, 140, 143, 151
Sima Qian, 154
situated action, 61
Six War Strategies 《六韬》, 125
Slack, Jennifer D., 81, 82, 84, 85
social conformity, 158, 163
social order and harmony, 11
social perspectives on technology, 60, 62
social stratum, 132, 146
social theories, 39, 53
Sohoo, 238
Somers, Robert, 19
Song Dynasty, 105, 124, 151, 153, 154, 155, 157, 158, 159, 160, 161, 162, 163, 164, 165
soot ink, 5, 119, 121, 122
sophistic rhetoric, 11, 73
Soviet Union, 44, 45, 46, 52, 96, 190
Spring and Autumn, 10, 15, 119, 129, 130, 134, 136
State Computer Industrial Administration, 180
State Council: Regulations on Protecting Computer Software, 200
State Economic Informationalization Joint Conference, 204
State Education Commission, 202, 205
State Information Department, 240
State Natural Science Foundation, 202
State Planning Commission, 202, 205
State Science Commission, 202
stone inscription, 129

subjectivity, 58, 84
substantive theory, 39, 41, 42, 43
Suchman, Lucy, 61
Sui Dynasty, 105, 164
Sui, Cindy, 211, 217
Sullivan, Patricia & Porter, James E., 63
Sun Bin's Art of War 《孙膑兵法》, 125
Sun Wu, 127
Sun Wu's Art of War 《孙子兵法》, 125
Sun Zi, 131
Sun Zi 《孙子》, 131
superego, 79
systematic management, 6, 82, 83, 86, 87
system-centered approach, 65

Taft Commission, 83
Tang Dynasty, 105, 113, 151, 152, 153, 155, 156, 170
Tao, 15, 17, 21, 22
Taoism, 10, 15-18, 21, 22, 23, 114, 131, 135, 145, 147, 163, 192, 193
Taoism, principle of relativity, 17
Taoism, view of language, 18
Taoist rhetoric, 18
technical artifact, 28, 29, 31, 32
technical elite, 78, 93, 95, 96, 98, 99, 132, 197, 198, 212, 214, 215, 221, 233
technical rationality, 26, 28, 31, 32, 51, 61, 77, 91, 92
technique: primary qualities, 54; secondary qualities, 54
techno-logic, 98
technological determinism, 39, 41, 42, 43, 46, 47, 50, 54, 59, 64
technology: as a process, 40; as a product, 40; cultural aspect, 26, 29, 31; definition, 25-27;

Index

four stages of conceptualization, 38-67; objective elements, 45; purpose, 64, 65; subjective elements, 45
technology as emergence, 37; cognitive understanding, 41, 47; exploratory interaction, 47; passive exposure, 38, 39; proactive control, 38, 53, 67
technology transfer, 4, 23, 28- 32, 33, 34, 35, 36, 37, 40, 41, 42, 43, 48, 49, 53, 55, 66, 68, 80, 87, 88, 89, 93, 102, 211
Temporary Administrative Measures on Internet Domain Registration in China, 206
The Analects 《论语》, 11, 135, 143
The Book of Change 《易经》, 10, 143
The Book of Han—The History of Art and Language, 150
The Book of History 《尚書》, 10, 143, 147
The Book of Mean 《中庸》, 10
The Book of Mencius 《孟子》, 11, 135, 143
The Book of Music, 147
The Book of Rhythms, 147
The Book of Rites 《礼记》, 10, 143
The Book of Songs 《诗经》, 10, 134, 143
The Doctrine of the Mean 《中庸》, 143
The Golden Medium 《中庸》, 10
The Great Learning 《大学》, 10, 143
The Han Chronicles 《汉书》, 147
The History by Zou 《左传》, 135
The Rhetoric of the Warring States 《战国策》, 135
The Sayings of Confucius 《论语》, 11

The Spring and Autumn Annals 《春秋》, 10, 143
theory of scientific and technological revolution (STR), 39, 41, 44, 45, 46, 47, 50, 52, 59
tortoise shell, 106, 107, 116, 117
Tsien, Tsuen-hsuin, 118, 120, 121, 122, 123, 124, 126, 127, 136, 138, 140, 153, 154
turtle shell, 5
Twitchett, Denis, 152, 156, 157, 158, 161, 162, 167
typesetting, 167

U.S. National Science Foundation, 203, 204
units of measurement, 128
Upper Cave Man, 112
user-centered approach, 64-66

Van Maanen, John, 34
vocation, 57, 58

Wang Chang Ling, 156
Wang Han, 156
Wang Shang Sheng Huo (Life on the Internet), 237
Wang Wei, 156
Wang Zhi Huan, 156
Wang, Haixia, 14, 18, 23, 72, 73, 194
Wang, Li, 106
Wang, Yeu-Farn, 208, 209, 231, 232
Wang, Zhaopeng, 170
Ware, James R., 11
Warring States Period, 10, 119, 120, 121, 123, 125, 127, 128, 129, 130, 131, 144, 147
weaponry inscription, 128
Weaver, Richard, 33, 34
Wen, Tiejun, 200
Western Han Dynasty, 126, 139, 141, 142, 143

Western ideologies, 192, 195, 196, 233
Western Zhou Dynasty, 110, 118, 134
Williams, Frederik & Gibson, David V, 28, 29, 88
Williams, Frederik & Gibson, David V., 35
Williams, Raymond, 25
Winner, Langdon, 26, 62
Winograd, Terry & Flores, Fenando, 61
wood slips, 5, 119
Wright, Arthur F., 19, 20
writing technologies in the West, 6
Wu Di Emperor, 108, 142
Wu, Yue, 228, 229

Xia Dyansty, 104, 112, 113, 118, 130, 132
Xia, Zhengnong, et al., 5, 8, 19, 105, 106, 154
Xian, 106, 137, 220
Xiaodao Lun 《笑道论》, 21, 22
Xiaotun, 106
Xie, Songxin & Xhang, Dingmin, 207
Xinjiang, 137

Xu Shen, 150
Xun Zi, 131, 135
Xun Zi 《荀子》, 135

Yan Zi 《晏子》, 125
Yanghe, Shandon Province, 8
Yangshao Culture, 106, 112
Yates, JoAnne, 6, 29, 35, 82, 83, 86
yi (义 justice), 12, 193
Yin-Yang School, 144, 145
Yu Liaozi 《尉缭子》, 125
Yu, Qiding, 146

Zhang, Jeff X. & Wang, Yan, 180, 187, 200
Zhao Ziyang, 200
Zhong Chong, 143
Zhong Xi Corporation, 239
Zhou Dynasty, 5, 21, 104, 109, 110, 112, 113, 114, 118, 119, 120, 122, 123, 124, 126, 127, 128, 129, 131, 132, 133, 134, 135, 141, 148, 149, 220
Zhu Rongji, 203, 207
Zhuang Zi, 15, 16, 17, 131
Zhuang Zi 《庄子》, 135
Zou Jiahua, 204, 205

About the Author

With a research interest mainly in the reciprocal relationship between writing technology development and cultural contexts, Baotong Gu's publications range from articles, reviews, and translations to four co-edited collections: *Content Management: Implications for Technical Communicators* (2008, a special issue *for Technical Communication Quarterly*); *Content Management: Bridging the Gap between Theory and Practice* (2009, Baywood); *Contemporary Western Rhetoric: Critical Methods and Paradigms* (1998, China Social Sciences Academy Press); and *Contemporary Western Rhetoric: Speech and Discourse Criticism* (1998, China Social Sciences Academy Press). Gu is an associate professor of English at Georgia State University.

www.ingramcontent.com/pod-product-compliance
Lightning Source LLC
Chambersburg PA
CBHW030132240426
43672CB00005B/107